International Zinc Conference and International Zinc Oxide Industry Conference 2018

Carlsbad, California, USA
11-14 February 2018

ISBN: 978-1-5108-5935-7

Printed from e-media with permission by:

Curran Associates, Inc.
57 Morehouse Lane
Red Hook, NY 12571

Some format issues inherent in the e-media version may also appear in this print version.

Printed by Curran Associates, Inc. (2018)

For permission requests, please contact International Zinc Association
at the address below.

International Zinc Association
2530 Meridian Parkway
Suite 115
Durham, NC 27713
USA

Phone: +1 919 361 4647

contact@zinc.org

Additional copies of this publication are available from:

Curran Associates, Inc.
57 Morehouse Lane
Red Hook, NY 12571 USA
Phone: 845-758-0400
Fax: 845-758-2633
Email: curran@proceedings.com
Web: www.proceedings.com

2018 International Zinc and Zinc Oxide Industry Conferences – Program

Zinc ...essential for life

Saturday, 10 February

3:00 – 5:00 PM	Registration Desk open

Sunday, 11 February

3:00 – 5:00 PM	Registration Desk open
6:00 – 7:30 PM	Welcome Reception

Monday, 12 February

07:30 AM – 01:00 PM	Registration Desk open	
08:45 AM – 09:00 AM	Introduction Stephen Wilkinson, International Zinc Association	N/A
09:00 AM – 10:30 AM	Keynote Speech: From Trump to Brexit - Why we need to expect the unexpected' Steve Richards, Political Journalist	N/A
10:30 AM – 11:00 AM	Investing in the Zinc Future Rodrigo Daud, CEO, American Zinc Recycling	1
11:00 AM – 11:30 PM	Plans, Projects and Priorities : Attempting to close the zinc mine supply gap Paul White, Secretary General, International Lead and Zinc, Copper and Nickel Study Groups	13

12:30 PM – 05:00 PM	Golf Tournament
01:00 PM – 03:00 PM	Tennis Tournament

Tuesday, 13 February

08:00 AM – 04:00 PM	Registration Desk open	
09:00 AM – 09:30 AM	Fireweed Zinc: Advancing the Macmillan Pass Zn-Pb-Ag Project Brandon Macdonald, CEO & Director, Fireweed Zinc	31
09:30 AM – 10:00 AM	World Class Zinc in the Developed World Tom Whelan, Arizona Mining	48
10:00 AM – 10:30 AM	How to maximize the market value of base and precious metal concentrates Boris Eykher, Co-Founder, Open Mineral	57
10:30 AM – 11:00 AM	Market Opportunities in Thermal Sprayed Zinc Frank Goodwin, International Zinc Association	66
11:00 AM – 11:30 AM	Developments in China Claire Hassall, Principal, CHR Metals	87

2018 International Zinc and Zinc Oxide Industry Conferences – Program

Zinc ...essential for life

Afternoon session joint with Zinc Oxide Industry Conference		
01:30 PM – 04:00 PM	Zinc Market Outlook including Panel Discussion Jonathan Leng, Wood MacKenzie Vivienne Lloyd, Macquarie Helen O'Cleary, CRU Duncan Hobbs and Luis Woolcott, Concord Resources Michael Cuoco, Wells Fargo	101
06:00 PM – 07:30 PM	Reception	

Wednesday, 14 February

8:00 AM – 2:30 PM	Registration Desk open	
9:00 AM – 9:30 AM	Zinc Oxide Market Overview Sean Davis, IHS Markit	156
9:30 AM – 10:00 AM	Updates on Zinc Nutrient Initiative Activities Andrew Green, IZA	164
10:00 AM – 10:30 AM	Recycling Small Batteries, Making a BIG Impact, an Overview of Alkaline Battery Recycling Shane Thompson, Battery Solutions	176
10:30 AM – 11:00 AM	Break	
11:00 AM – 11:30 AM	Regulatory Trends Impacting the Tire Industry in 2018 Sarah Amick, US TMA	187
11:30 AM – 12:00 PM	Reduce Production Costs and Increase Productivity using Energy Efficient Crucibles Mike Zeno, Foseco	200
12:00 PM – 1:30 PM	Lunch	
1:30 PM – 2:00 PM	Update: EU Regulatory Issues and Beyond Christine Spirlet, IZA	212
2:00 PM – 2:30 PM	Trajectory of Environmental Regulations in the Americas Eric Van Genderen, IZA	225
3:00 PM – 4:30 PM	Zinc Oxide Committee Meeting	

SAVE THE DATE

February 24 - 27, 2019

2019 International Zinc Conference & Zinc Oxide Industry Conference

Westin Kierland Resort
Scottsdale, Arizona

SPONSORS

GLENCORE

ocean PARTNERS

CLIVEDEN TRADING AG

OPENMINERAL

SUMMIT MINERALS GmbH

TRAXYS

Richker Metals Inc.

HOW TO DOWNLOAD AND USE THE MOBILE EVENT APP

STEP 1

DOWNLOAD THE APP

Search for ZINC EVENTS in the Apple, Android, BlackBerry or Windows App store and download the App for free on your smart phone or tablet device.

STEP 2

LOGIN TO THE EVENT

To login, click on the 2018 International Zinc & Zinc Oxide Industry Conferences event – then enter your email address and create your personal password. Allow for push notifications so you can receive instant updates.

STEP 3

BEGIN USING THE APP

Access the main menu to:
- Create your schedule
- View a list of attendees & contact information
- See a list of speakers
- Send private messages
- Manage your own data

Access the App via a web browser from your computer and download the presentations after the conference:

https://eventmobi.com/2018_intl-zinc-and-zno_conference/

SATURDAY, 10 FEBRUARY

3:00 - 5:00 PM
Registration Desk Open - The Nest Foyer

SUNDAY, 11 FEBRUARY

3:00 - 5:00 PM
Registration Desk Open - The Nest Foyer

6:00 - 7:30 PM
WELCOME RECEPTION - Palm Courtyard

MONDAY, 12 FEBRUARY

7:00 - 8:30 AM
BREAKFAST - Aviara Salon & Foyer, Courtyard

7:30 AM - 12:00 PM
Registration Desk Open - Grand Foyer

8:45 - 11:30 AM
General Session: International Zinc Conference
Grand Ballroom

Introduction
Stephen Wilkinson, IZA

KEYNOTE SPEECH
From Trump to Brexit - Why we need to expect the unexpected
Steve Richards, Political Journalist

Investing in the Zinc Future
Rodrigo Daud, American Zinc Recycling

Plans, Projects and Priorities: Attempting to close the Zinc Mine Supply Gap
Paul White, International Lead and Zinc, Copper and Nickel Study Groups

NOTE: BADGES REQUIRED AT ALL TIMES.

TUESDAY, 13 FEBRUARY

7:00 - 8:30 AM
BREAKFAST - Aviara Salon & Foyer, Courtyard

8:00 AM - 4:00 PM
Registration Desk Open - Grand Foyer

9:00 - 11:30 AM
General Session: International Zinc Conference
Grand Ballroom

Fireweed Zinc: Advancing the Macmillan Pass Zn-Pb-Ag Project
Brandon Macdonald, Fireweed Zinc

World Class Zinc in the Developed World
Tom Whelan, Arizona Mining

How to maximize the market value of base and precious metal concentrates
Boris Eykher, Open Mineral

Market Opportunities in Thermal Sprayed Zinc
Frank Goodwin, IZA

Developments in China
Claire Hassall, CHR Metals

1:30 - 4:00 PM
Joint Session
Int'l Zinc & Zinc Oxide Industry Conferences
Grand Ballroom

ZINC MARKET OUTLOOK
Jonathan Leng, Wood MacKenzie
Vivienne Lloyd, Macquarie
Helen O'Cleary, CRU
Duncan Hobbs and Luis Woolcott, Concord Resources
Michael Cuoco, Wells Fargo

6:00 - 7:30 PM
RECEPTION - Avalon Salon & Foyer, Terrace

WEDNESDAY, 14 FEBRUARY

8:00 AM - 3:00 PM
Registration Desk Open - Grand Foyer

9:00 AM - 12:00 PM
General Session:
Int'l Zinc Oxide Industry Conference
Grand Ballroom

Zinc Oxide Market Overview
Sean Davis, IHS Markit

Updates on Zinc Nutrient Initiative Activities
Andrew Green, IZA

Recycling Small Batteries, Making a BIG Impact, an Overview of Alkaline Battery Recycling
Shane Thompson, Battery Solutions

Regulatory Trends Impacting the Tire Industry in 2018
Sarah Amick, US TMA

Reduce Production Costs and Increase Productivity using Energy Efficient Crucibles
Michael Zeno, Foseco

12:00 PM - 1:30 PM
LUNCH - Grand Ballroom D

1:30 PM - 2:30 PM
General Session:
Int'l Zinc Oxide Industry Conference
Grand Ballroom

Update: EU Regulatory Issues and Beyond
Christine Spirlet, IZA

Trajectory of Environmental Regulations in the Americas
Eric Van Genderen, IZA

3:00 PM - 4:30 PM
Zinc Oxide Committee Meeting - Aviara Salon

Investing In the Zinc Future

Rodrigo Daud, CEO – American Zinc Recycling
Chair – International Zinc Association

International Zinc Association

International Zinc Association

Mission Statement

Representing the Zinc Industry globally, to sustainably grow markets and maintain the Industry's license to operate through effectively managed initiatives in research & development, technology transfer and communication of the value of zinc

International Zinc Association

Investments in Technology & Market Development

International Zinc Association

Galvanized Autobody Partnership

GAP

Recent Findings

- Galvanizability of 3G steels
- Keys to reducing zinc coating weight variability defined
- Toolbox developed for bath analysis and management
- Zn-coated tailor welded blank laser welding procedures and properties defined

International Zinc Association

United States Steel Corporation

United States Steel Corporation and Kobe Steel Announce New Investment in Advanced High-Strength Steel Capabilities

NEWSROOM ITEM | SEPTEMBER 25, 2017

"New Audi A8 switches back to steel after Experimenting with aluminum...new model will be released in 2018"

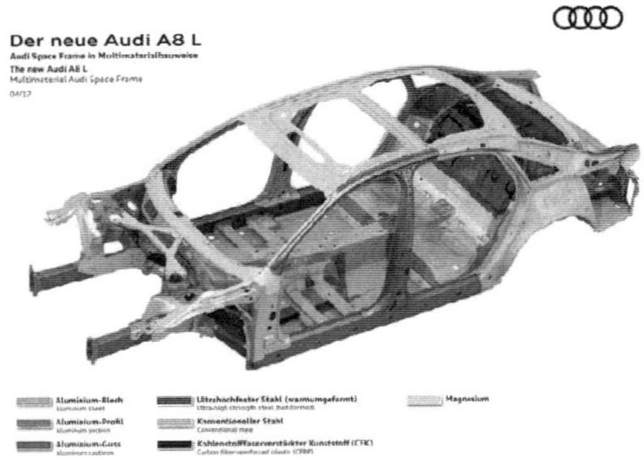

Zinc Nutrient Initiative

- Seal of Approval launched in Brazil to improve zinc fertilizer sourcing quality
- Programs in Democratic Republic of the Congo (DRC) with funding from Lundin Foundation & MMG successful
- New initiative being launched with World Bank to expand impact

THE WORLD BANK
IBRD • IDA

Bloomberg Markets Tech Pursuits Politics Opinion Businessweek Sign In Subscribe

Nutrient-Poor Farms Get a Vitamin Boost From Zinc Mines

By Danielle Bochove

April 17, 2017, 7:00 PM EDT *Updated on* April 18, 2017, 5:50 AM EDT

- **IHS Markit – 270,000 tpy market in 2017**
- **Potential is for 900,000 tpy**

International Zinc Association

Continuous Galvanized Rebar (CGR)

- Startup of first North American CGR Line -
- Two new lines likely in 2018: US and India
- CGR approved for use in VT and FL, others evaluating
- ASTM and AASHTO include CGR specifications

International Zinc Association

Zn Cast

VOL 1, NO 1 - SEPTEMBER 19, 2017

Dirndls Captivate With High-Quality Zinc Details

Dirndls are a style of traditional dress worn in Austria, Bavaria and South Tyrol, based on the work clothing of Alps peasants. For many years dirndls were a nostalgic costume dress reserved for festive occasions such as the Munich Oktoberfest. But dirndls have enjoyed a renaissance of late, even reappearing as hip mainstream fashion. Rarely alike, the appeal of dirndles is their affinity for individualization and detail. In addition to rich fabrics, embroidery and appliqués; dirndles often adorned with matching buttons, hooks, eyelets and jewelry - which are commonly made of zinc.

Read More

Winners of 2017 INTERZINC Design Challenge

The INTERZINC Design Challenge, the World's largest student zinc die casting design competition, announced its 2016 - 2017 winners. Entries were carefully evaluated by an internationally recognized panel of zinc die casting experts based on the appropriateness of the solution to the design problem, including but not limited to: quality, originality, thoroughness and commercialization potential.

The Design Challenge was established to create awareness for good zinc die casting design practices and principles amongst engineering and design students. The main component of the competition is to demonstrate the advantages of using zinc die castings. The theme for this year's competition was Commercial Kitchen Products.

Read More

Ulfberht Knife

Bryan Varga of the University of Oregon made use of zinc's natural antibacterial properties to create a general purpose cooking knife with a zinc handle and stainless steel blade.

Save-A-Shot

Designed by Garen Gibbs of Purdue University. Save-A-Shot resolves the issue of commonly breaking espresso shot glasses, while also reducing clutter around espresso machines, increasing productivity.

Zinc Faucet

Although zinc is already used in faucets, Cole Sippel of the University of Wisconsin – Stout, brought a new imaginative approach to a one hand control faucet using electronic controls on the nozzle.

Learn More About the Winning Entries

CHICAGO

June 4-6, 2018

Chicago O'Hare Airport Sheraton Hotel

International Zinc Die Casting Conference 2018

Die Casting

- 1st International Die-Casting Conference and Design Competition in NA - Chicago
- Website updated with French, German and Spanish translations. E-case book updated
- Brochure "Engineering in Zinc – Today's answer" translated into German and French, Chinese and Spanish underway
- German group of die casters, alloy and machine producers established for cooperative MD work

International Zinc Association

Recent developments in zinc die cast applications reaching market place

New Nespresso machine using zinc die cast

Zinc die-castings find growing Use in automotive sector

PORSCHE

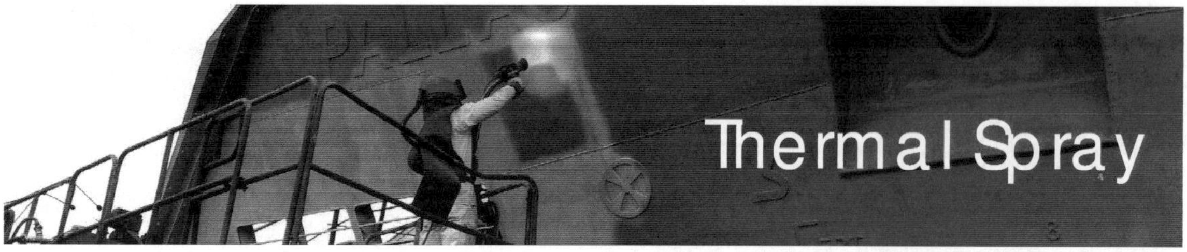

Thermal Spray

- One-year inspection of transport ship "Rasmus" in Sweden
- New brochure and website developed
- Siemen's now specifying thermal spray for all off-shore wind towers
- Focus now on ships, bridges and other applications

International Zinc Association

Renewable Energy Market Development For Galvanized Steel

- Identify major project developers PV, CSP, solar thermal sectors
- Raise awareness and provide technical assistance through engineering & cost data, inspection reports, brochures, website, etc.
- Develop case studies of successful projects, preferably with demonstrable positive maintenance & inspection results on durability

India Highlights

- New BIS stds. for construction & auto steels
- Thermal spray of railway tracks
- 1st integrated Zn die casting-electroplating plant opened
- Focus on zinc use in new "smart city" concept
- New IZA India website & published articles

International Zinc Association

China Highlights

- China GAP: two domestic automakers now use Zn-coated advanced steels
- Galvanized rebar standards achieved
- Focus on zinc usage in new smart city concept

International Zinc Association

New Opportunities in Batteries

Mining Journal · Global Edition

Market Data Webinar NEWSLETTERS TRIAL SUBSCRIBE LOGIN

EXPLORATION/DEVELOPMENT | FINANCE | COMMODITIES | VIEWPOINT | METS INVESTOR | MINE RISK MANAGEMENT | FUTURE OF MINING | ALL SECTIONS

Commodities > Base-metals

"Zinc metal demand (for) battery storage would be 19.8Mt of metal"

A Canadian commodity research firm has said older zinc battery technology could be a more efficient grid-scale energy storage solution than the much-hyped lithium equivalent, possibly boosting demand for a metal already at a 10-year price high.

Staff Reporter

New Bose Earbud using Zn-Ag batteries

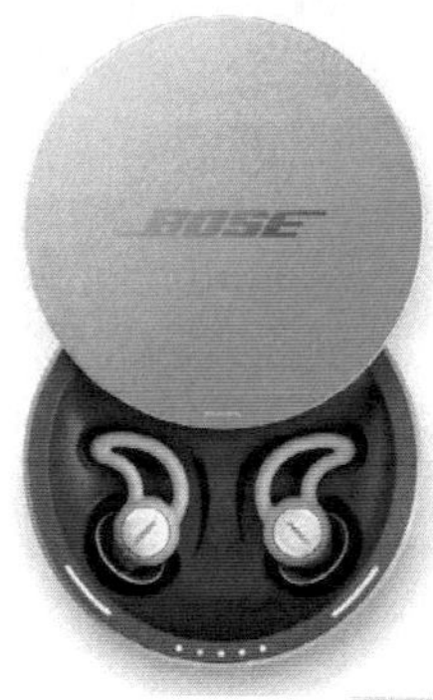

Investing in License to Operate... Environment, Health and Sustainability

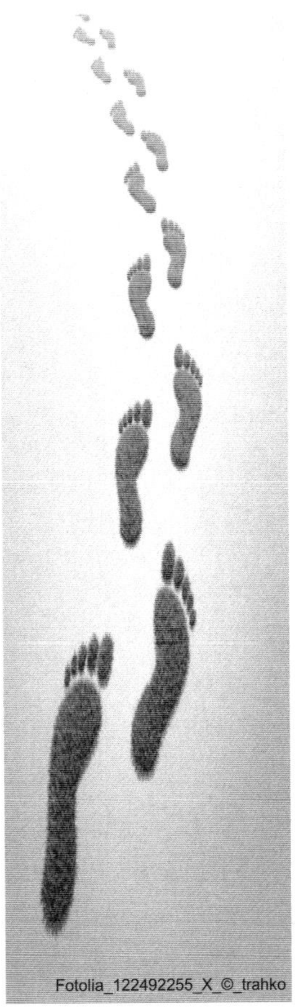

Fotolia_122492255_X_©_trahko

Sustainability

- Promote zinc products and market access
- Develop life cycle awareness related to zinc production, downstream use, and waste management
- Use scientifically sound and cost-effective strategies to resolve emerging issues

International Zinc Association

Sustainability

Zinc: Contributing to a Circular Economy

Unlike a traditional linear economy (extract, make, use, dispose), in a circular economy, each new object is designed for maximum longevity and future reuse and, at the end of its life, becomes a potential resource, rather than an item of waste for disposal.

Circular economy encompasses more than the production and consumption of goods and services; it aims to disconnect economic growth from the depletion of natural resources.

Given its many attributes, including being essential, versatile, durable, and infinitely recyclable, zinc contributes to a circular economy in numerous ways.

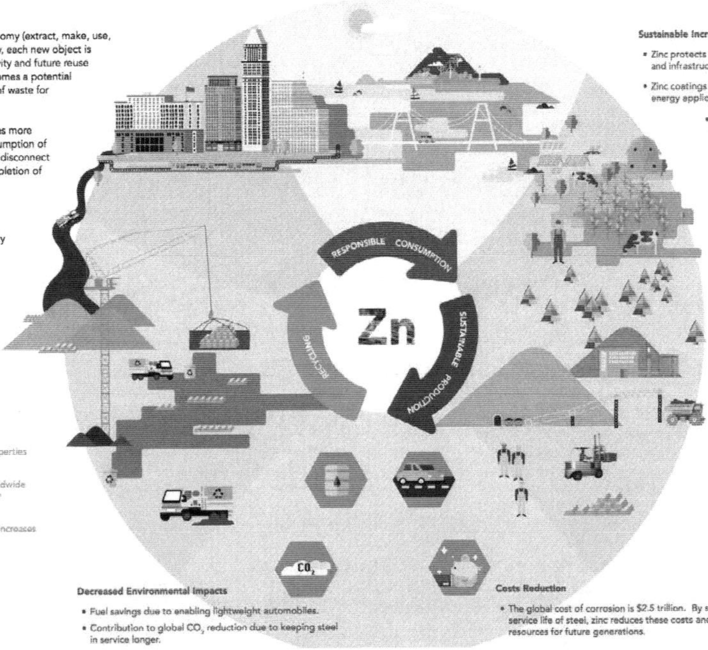

Sustainable Increased Service Life

- Zinc protects steel from corrosion in transportation and infrastructure.
- Zinc coatings extend the service life of solar and wind energy applications.
- The service life of many zinc products exceeds 50 years:
  - Zinc metal roofs and facades have a product life of 100 years or more.
  - Hot-dip galvanizing provides protection from 10 to 170 years.

Sustainable Improved Profitability

- In addition to improving nutritional value, zinc increases crop yields which improves farmer's income.

Abundant Natural Resources

- Zinc is the 24th most abundant natural element.
- Zinc is essential for all living organisms.

Sustainable Production

- IZA members have adopted a set of commitments to sustainability as set out in the Sustainability Charter.[1]
- Increased energy efficiency of zinc production. Since 2005, zinc producers have achieved a 24% reduction in primary energy needs.

Efficient Zinc Recycling

- Zinc keeps its physical properties across its life cycle.
- 30% of zinc produced worldwide originates from recycled or secondary zinc.
- The level of zinc recycling increases each year.

Decreased Environmental Impacts

- Fuel savings due to enabling lightweight automobiles.
- Contribution to global CO_2 reduction due to keeping steel in service longer.

Costs Reduction

- The global cost of corrosion is $2.5 trillion. By significantly increasing the service life of steel, zinc reduces these costs and protects valuable steel resources for future generations.

[1] http://www.zinc.org/sustainability

Zinc Product Stewardship

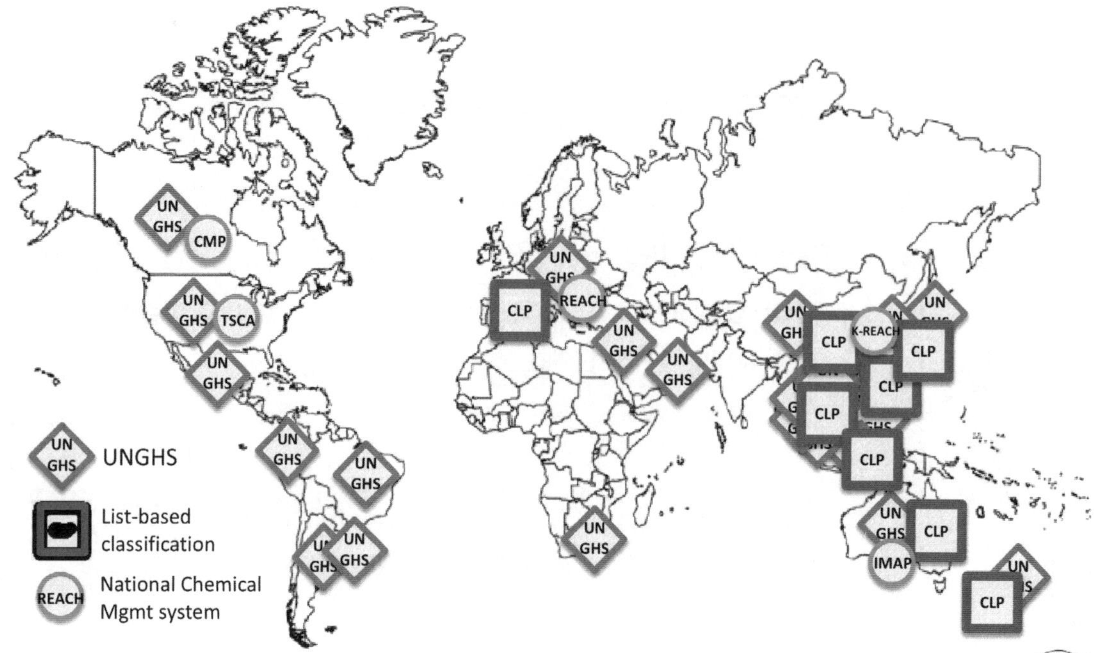

Regulatory Affairs

- IZA provides global and regional responses to regulatory challenges - e.g. REACH
- Zinc taken off list for EU Water Framework Directive
- Official partnership with USEPA on new regulatory water quality initiative
- California bills preventing use of ZnO in tires stopped in 2016 & 2017
- Norwegian authorities convinced to drop anti-zinc recommendation in construction

Health & Corporate Social Responsibility

- IZA taking global lead in advocating programs aimed at addressing zinc deficiency
- Initiative launched in Mexico with Penoles, Zinc Nacional and UNICEF
- Zinc nutrition programs launched in Africa

ZINC Saves Kids®

An International Zinc Association (IZA) initiative in support of UNICEF

Communications

- IZA supports flagship site zinc.org as well as > 12 individual market development sites for regional programs and major end-uses

Health

Die Casting

Galvanized Sheet

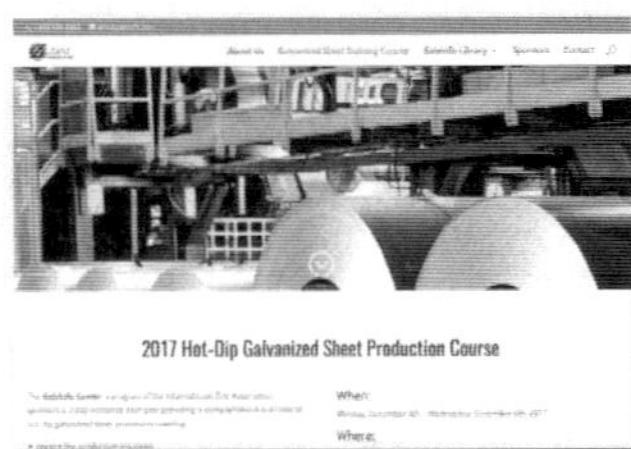

Galvanized Rebar

Zinc Fertilizers

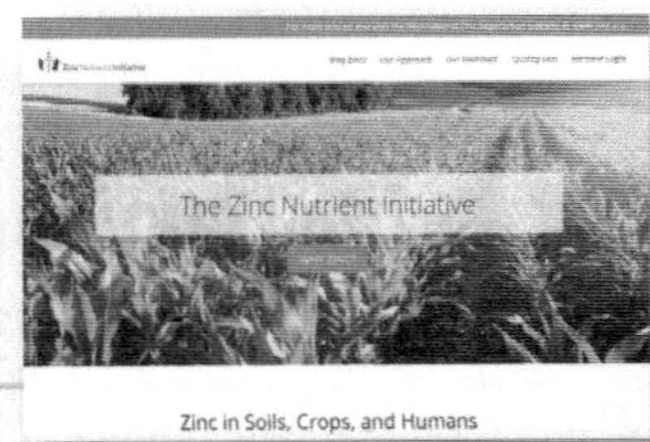

Thermal Spray

Conferences

- **Plus 7 additional regional/application conferences**

Summary

- High potential programs at various stages of development.

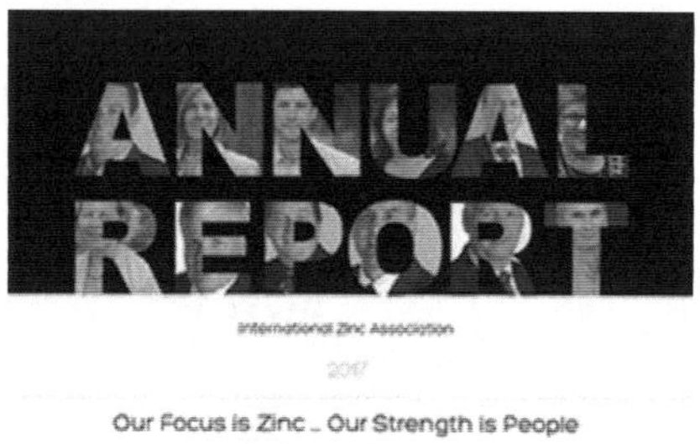

- Investing in IZA and zinc projects having significant impact on market protection and growth

Our Focus is Zinc: Our Strength is People

International Zinc Conference 2018
Carlsbad, 12 February 2018

Plans, Projects and Priorities - Attempting to Close the Gap in Global Zinc Mine Supply

Paul White
Secretary General
International Copper, Nickel and Lead and Zinc Study Groups

Presentation Outline

- ILZSG Overview
- Zinc Mine Supply General Trends
- Cutbacks and Closures
- Recent Openings and Committed Additions
- Longer Term Possibilties
- Conclusions

ILZSG Overview

- Intergovernmental organization set up within the UN system
- Significant level of industry representation
- Established by UN in 1959 in New York
- Moved to London in 1977
- From start of 2006 ILZSG, ICSG & INSG co-located in Lisbon, Portugal

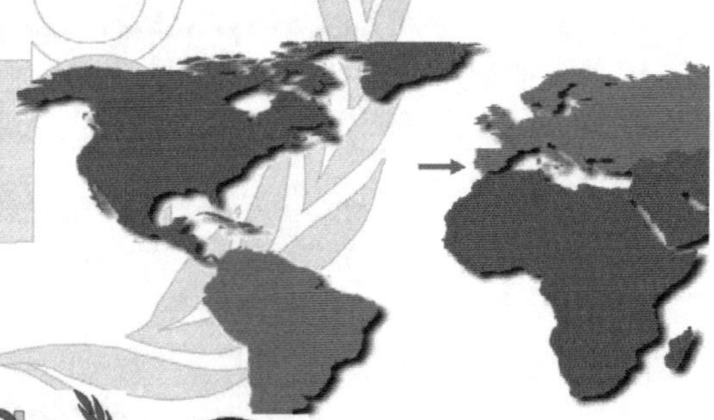

www.icsg.org ILZSG www.insg.org

ILZSG Membership

- Membership open to any country involved in lead and/or zinc production, usage, or trade.
- 28 members (>85% of global lead/zinc industry):

Australia	India	Morocco	Russian Fed
Belgium	Iran	Namibia	Serbia
Brazil	Ireland	Netherlands	Sweden
Bulgaria	Italy	Norway	Turkey
China	Japan	Peru	United States
Finland	Korea Rep.	Poland	European Community
France	Mexico	Portugal	
Germany			

ILZSG Overview – Work of the Group

- **Promote Market Transparency**
  - Closely monitor production, consumption, prices, stocks, trade flows and market balances
  - Reports and directories
- **Facilitate Co-operation Between Government and Industry**
  - Twice yearly meetings
  - Special conferences/seminars
- **In-depth Research into Issues of Interest/Concern to Members**
  - Environmental legislation
  - Economic developments

www.ilzsg.org International Lead and Zinc Study Group 5

Presentation Outline

- ILZSG Overview
- Zinc Mine Supply General Trends
- Cutbacks and Closures
- Recent Openings and Committed Additions
- Longer Term Possibilties
- Conclusions

www.ilzsg.org International Lead and Zinc Study Group 6

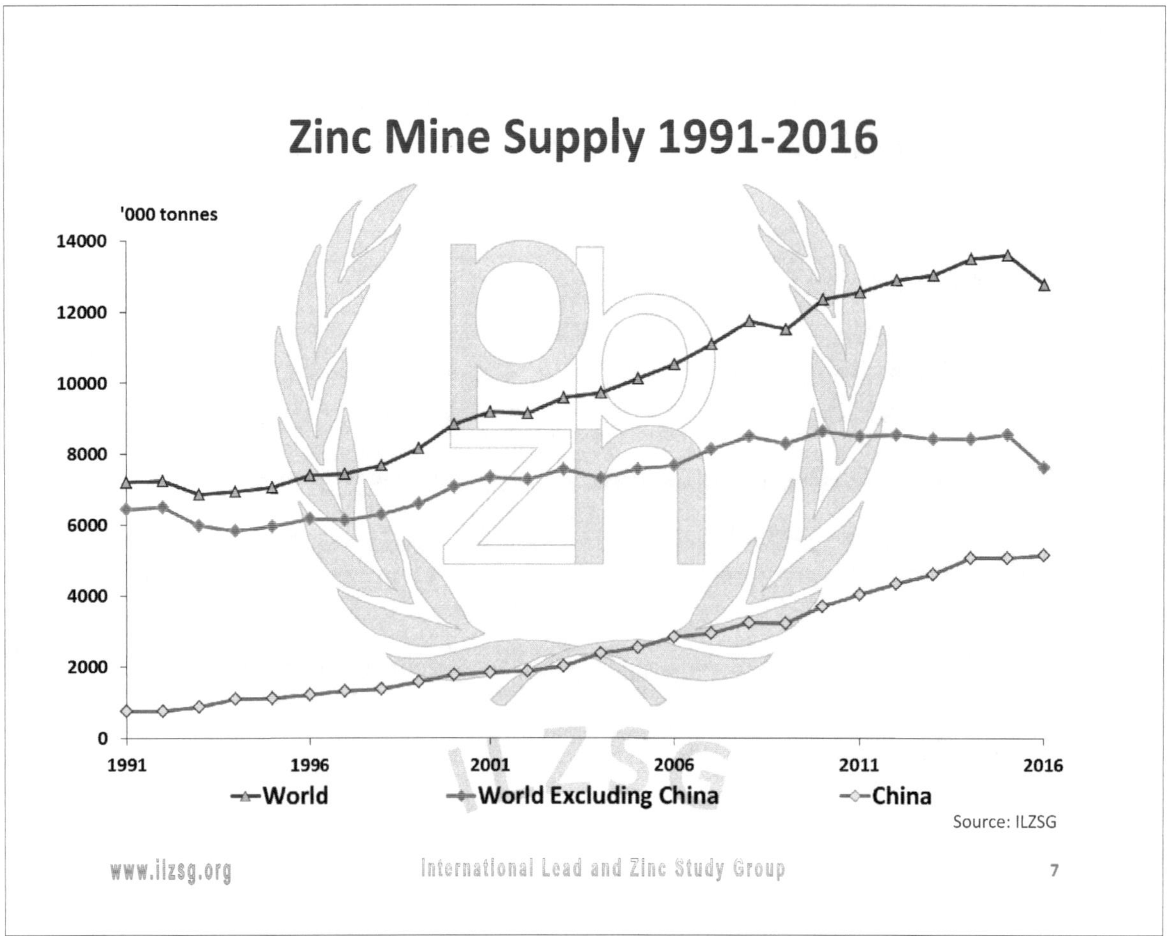

Zinc Mine Supply 1991-2016
'000 tonnes
14000
12000
10000
8000
6000
4000
2000
0
1991
1996
2001
2006
2011
2016
World
World Excluding China
China
Source: ILZSG
www.ilzsg.org
International Lead and Zinc Study Group
7

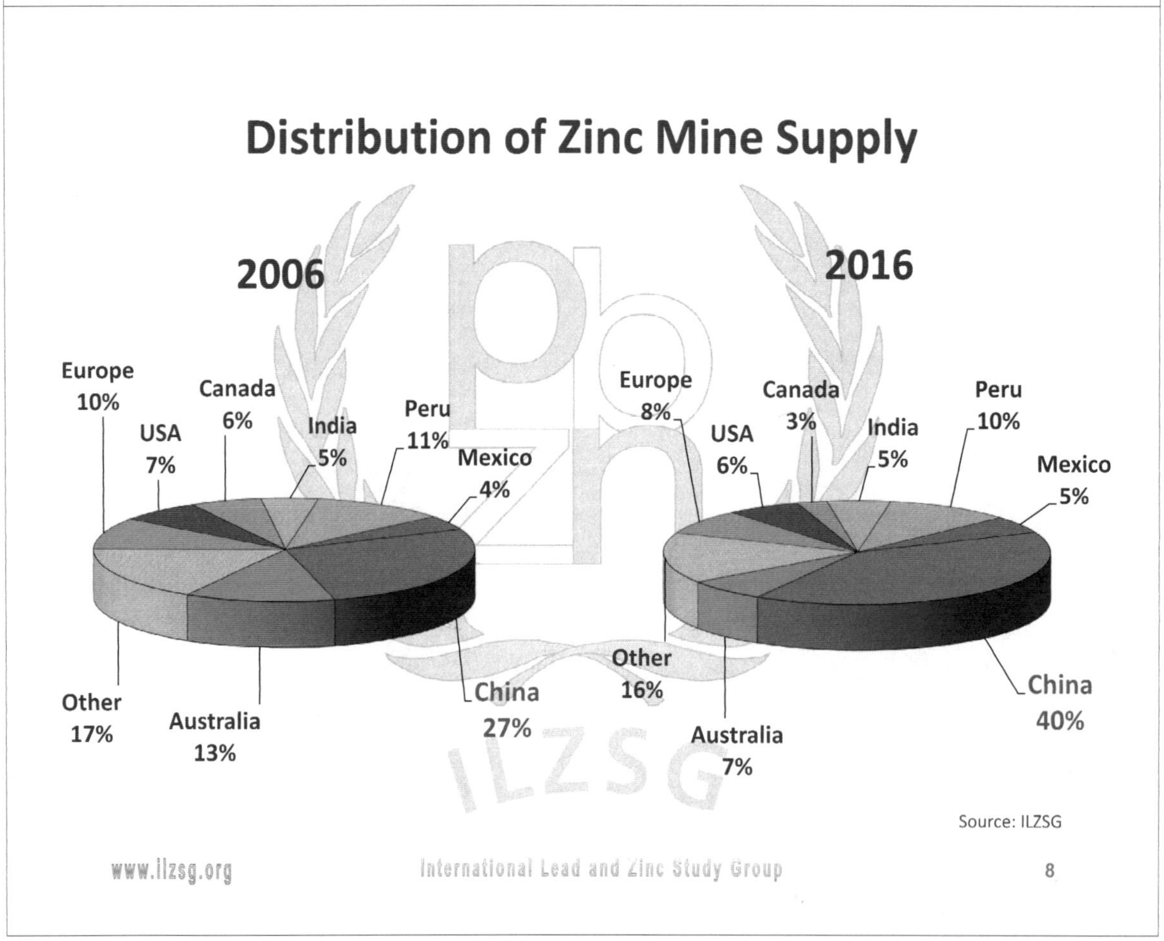

Distribution of Zinc Mine Supply
2006
Europe 10%
Canada 6%
USA 7%
India 5%
Peru 11%
Mexico 4%
China 27%
Australia 13%
Other 17%
2016
Europe 8%
Canada 3%
USA 6%
India 5%
Peru 10%
Mexico 5%
China 40%
Australia 7%
Other 16%
Source: ILZSG
www.ilzsg.org
International Lead and Zinc Study Group
8

Strong Growth in Indian Zinc Mine Output

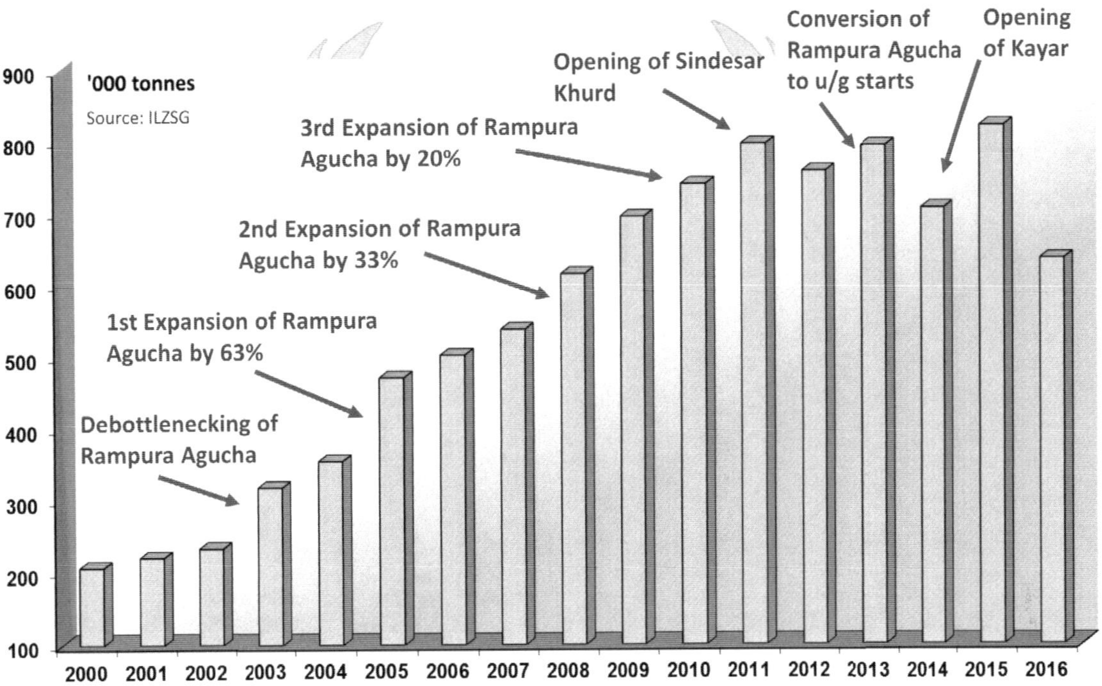

www.ilzsg.org International Lead and Zinc Study Group 9

Turkish Zinc Mine Output has also Risen

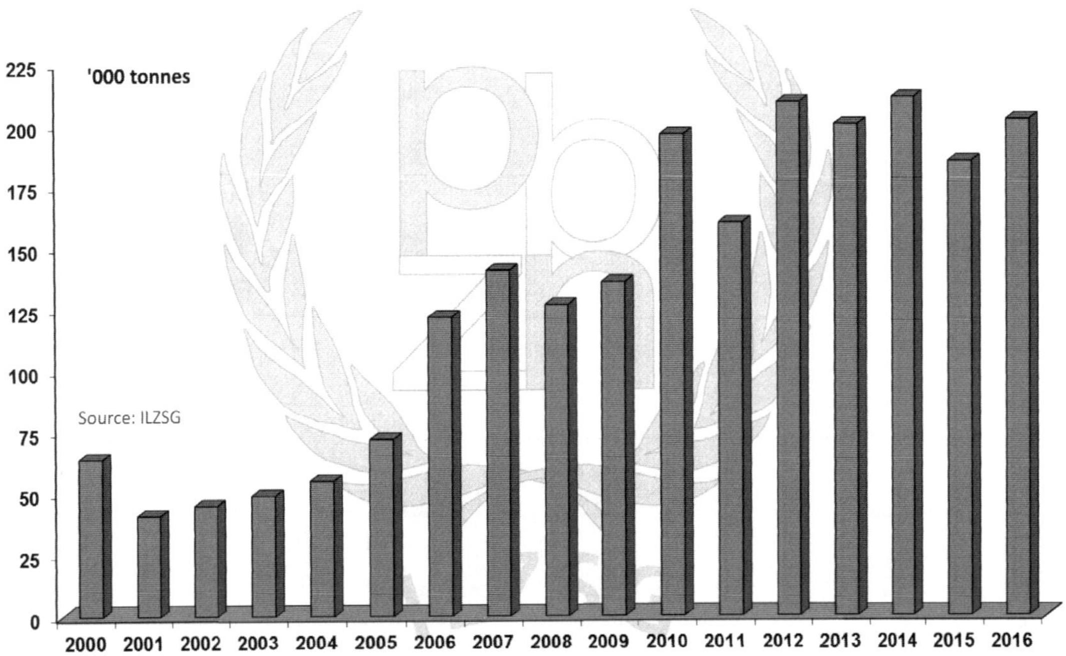

www.ilzsg.org International Lead and Zinc Study Group 10

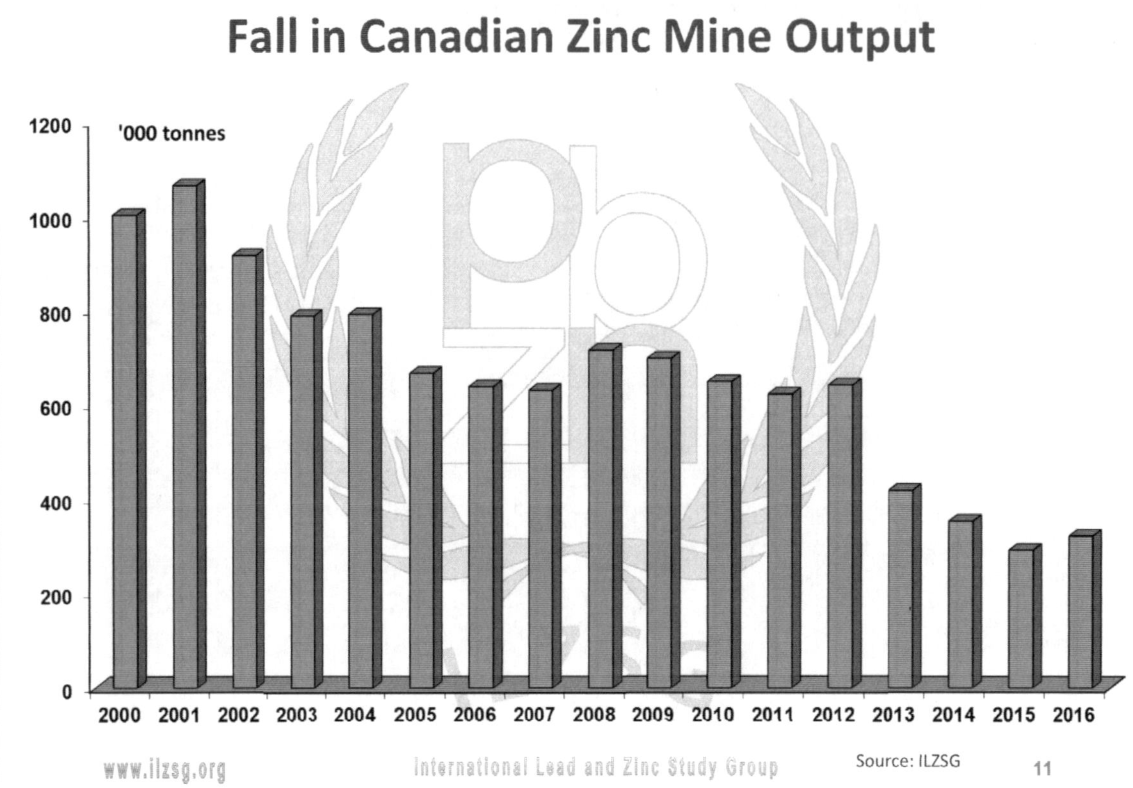

Fall in Canadian Zinc Mine Output
'000 tonnes
1200
1000
800
600
400
200
0
2000 2001 2002 2003 2004 2005 2006 2007 2008 2009 2010 2011 2012 2013 2014 2015 2016
www.ilzsg.org
International Lead and Zinc Study Group
Source: ILZSG
11

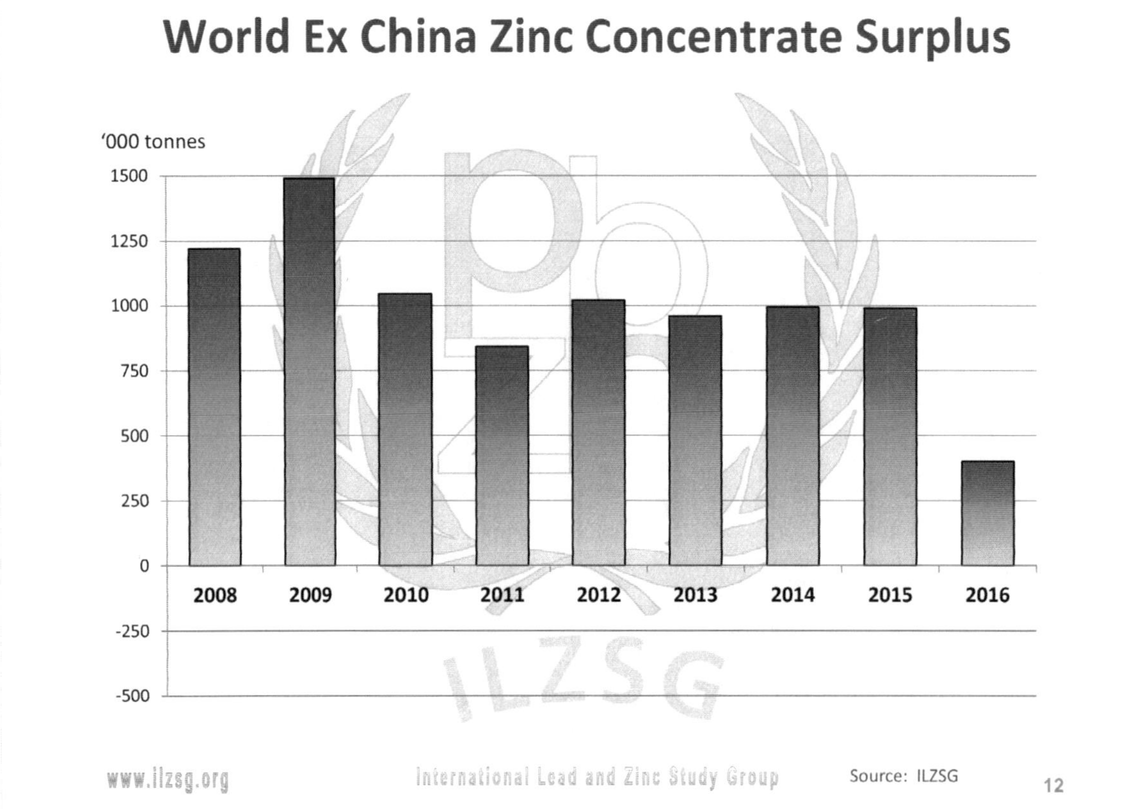

World Ex China Zinc Concentrate Surplus
'000 tonnes
1500
1250
1000
750
500
250
0
-250
-500
2008 2009 2010 2011 2012 2013 2014 2015 2016
ILZSG
www.ilzsg.org
International Lead and Zinc Study Group
Source: ILZSG
12

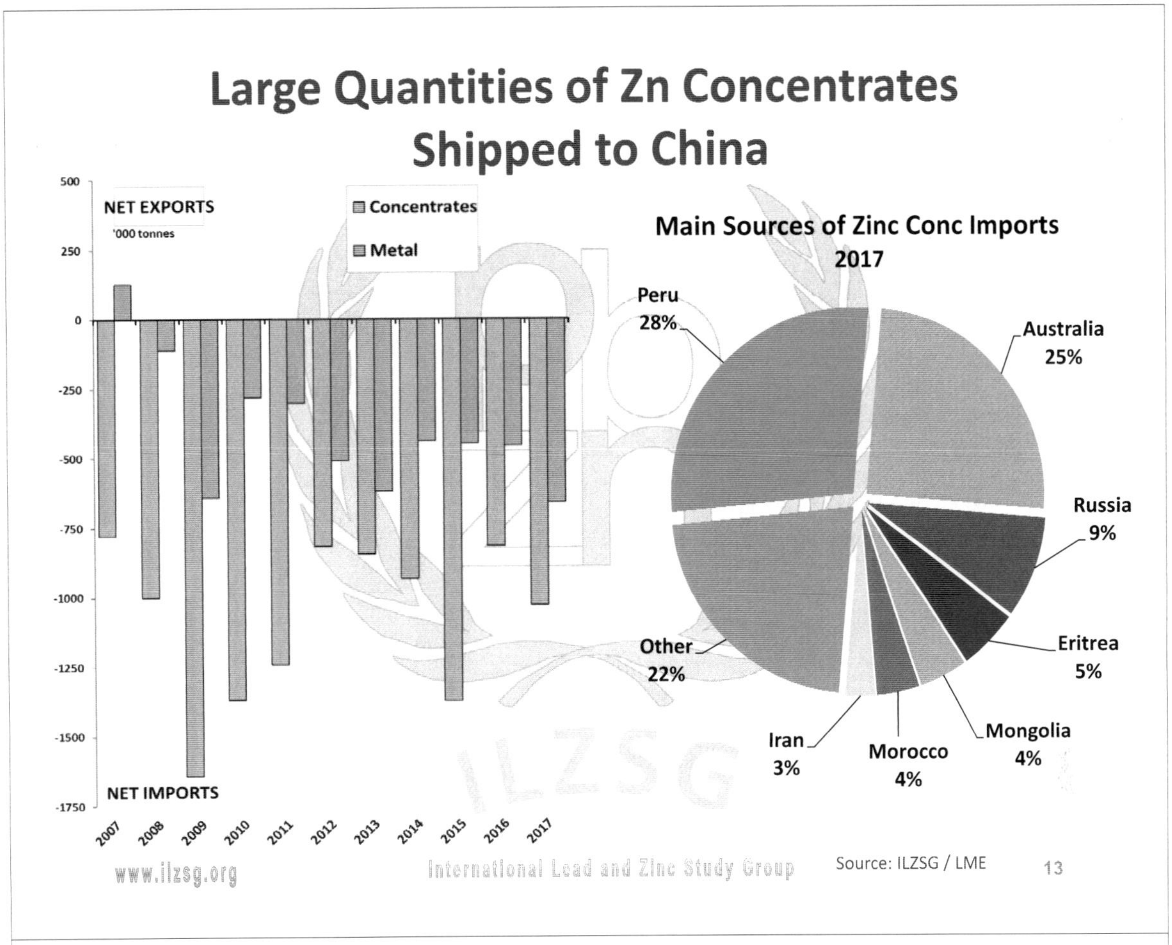
Large Quantities of Zn Concentrates Shipped to China
NET EXPORTS
'000 tonnes
Concentrates
Metal
500
250
0
-250
-500
-750
-1000
-1250
-1500
-1750
NET IMPORTS
2007
2008
2009
2010
2011
2012
2013
2014
2015
2016
2017
Main Sources of Zinc Conc Imports 2017
Peru 28%
Australia 25%
Russia 9%
Eritrea 5%
Mongolia 4%
Morocco 4%
Iran 3%
Other 22%
www.ilzsg.org
International Lead and Zinc Study Group
Source: ILZSG / LME
13

Weight of China in Selected Zn Conc. Exporters 2007-2017
Australia
Other 60%
China 40%
Peru
Other 74%
China 26%
Canada
Other 86%
China 14%
Morocco
Other 60%
China 40%
Mongolia
China 100%
Russia
China 63%
Other 37%
www.ilzsg.org
International Lead and Zinc Study Group
Source: ILZSG
14

Presentation Outline

- ILZSG Overview
- Zinc Mine Supply General Trends
- Cutbacks and Closures
- Recent Openings and Committed Additions
- Longer Term Possibilties
- Conclusions

www.ilzsg.org International Lead and Zinc Study Group 15

World Zinc Mine Capacity Changes 2001 – 2017

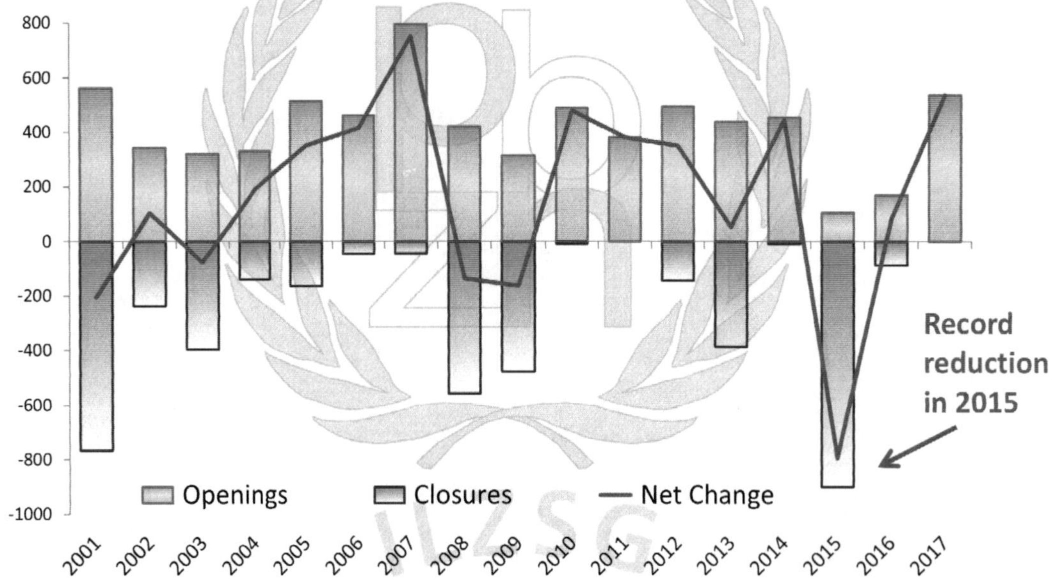

www.ilzsg.org International Lead and Zinc Study Group Source: ILZSG 16

Selected Zinc Mine Closures in 2015

Mine	Closed Annual Capacity (Zinc contained)
Century, Australia	-500,000t
Duck Pond, Canada	-34,000t
Myra Falls, Canada	-27,000t
Wolverine, Canada	-53,000t
Lisheen, Irelend	-175,000t
Campo Morado, Mexico	-10,000t
Naica, Mexico	-25,000t
Pallca, Peru	-15,000t
Mid-Tennessee (Gordonsville), USA	-60,000t

Source: New Mines and Smelters 2016 Report ,ILZSG

Zinc Mine Supply Cutbacks in 2016

- Glencore reduced zinc mine output by 500kt/yr in Australia, Kazakhstan and Peru
- Reduced output at HZL's Rampura Agucha operation in India due to waste stripping
- CBH Resources and Perilya reduced production at Endeavour and Broken Hill mines in Australia
- Operations at MMG's Golden Grove operation in Australia moved onto a campaign basis
- Suspension of output at Al Masane in Saudi Arabia

DPR Korea Zinc exports to China
2004 – 2017

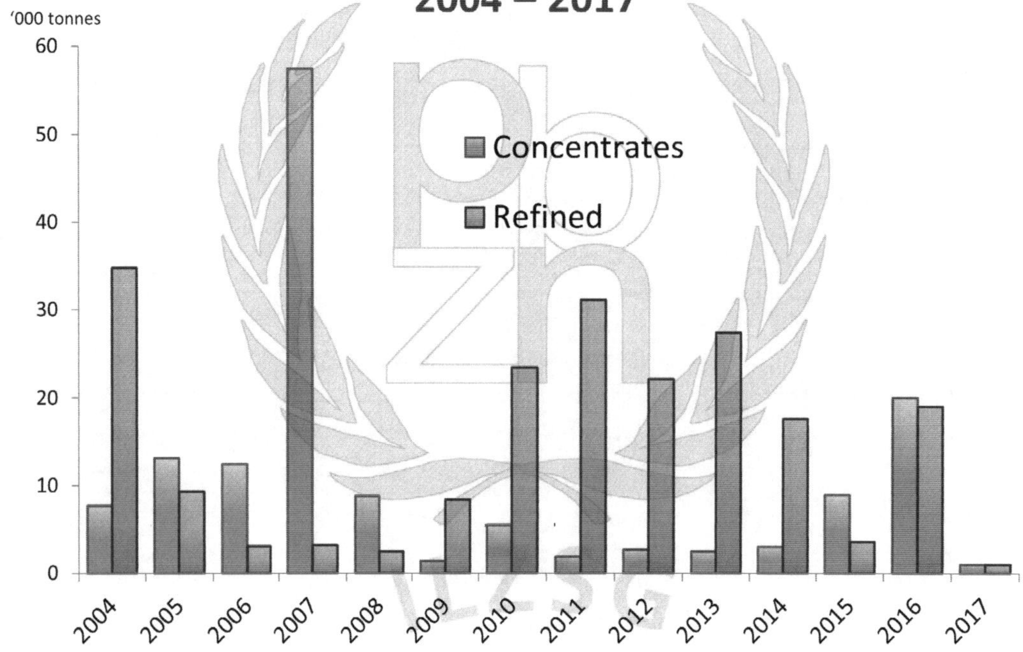

Source: ILZSG

World Zinc Mine Supply 2011 - 2016

Steep Fall in 2016

- **2016 - Global -6.2%**
  - China 1.5%
  - Ex China -10.7%

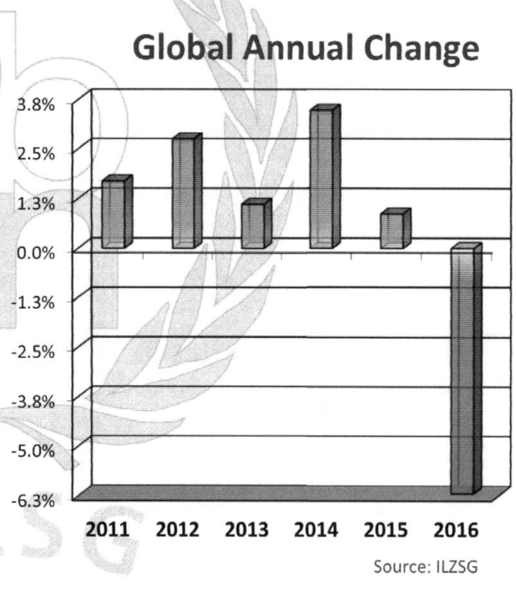

Source: ILZSG

Presentation Outline

- ILZSG Overview
- Zinc Mine Supply General Trends
- Cutbacks and Closures
- Recent Openings and Committed Additions
- Longer Term Possibilties
- Conclusions

Selected Additions to Zinc Mine Capacity in 2017

Source: New Mines and Smelters 2017 and 2018 Reports, ILZSG

Mine	Annual Capacity (Zn contained)	Opened
Dugald River, Australia	170,000t	Nov
Thalanga, Australia *(reopening)*	21,000t	Sept
Silvertip, Canada	15,000t	Jan
Castellanos, Cuba	50,000t	Dec
Olympias, Greece	21,000t	June
Campo Morado, Mexico *(reopening)*	20,000t	Nov
Marcona, Peru *(tailings)*	30,000t	May
Zinkgruven, Sweden *(expansion)*	10,000t	May
Mid-Tennessee, US *(reopening)*	50,000t	April

Selected Committed Additions to Zinc Mine Capacity 2018-2019

Source: New Mines and Smelters 2018 Reports ILZSG

Mine	Annual Capacity (Zn contained)	Scheduled Opening
Century Tailings, Australia *(tailings)*	264,000t	Q3 2018
Lady Loretta, Australia *(reopening)*	160,000t	2018/2019
Halfmile Lake, Canada	55,000t	Q3 2018
Iscaycruz, Peru *(reopening)*	80,000t	2019
Neves Corvo, Portugal *(expansion)*	75,000t	Q4 2019
Rey de Plata, Mexico *(reopening)*	34,000t	Q4 2018
Shalkiya – stage 1, Kazakhstan *(reopening)*	50,000t	2019
Gamsberg, South Africa	250,000t	Q2 2018
Empire State, US *(reopening)*	60,000t	Q1 2018

Gamsberg Project, South Africa

- Initially intention was to build mine/refinery complex
- Ore has high Mn content
- Development of mine started in Aug 2015
- Zinc capacity of 250kt/yr
- Majority of concs to be treated at Skorpion in Namibia

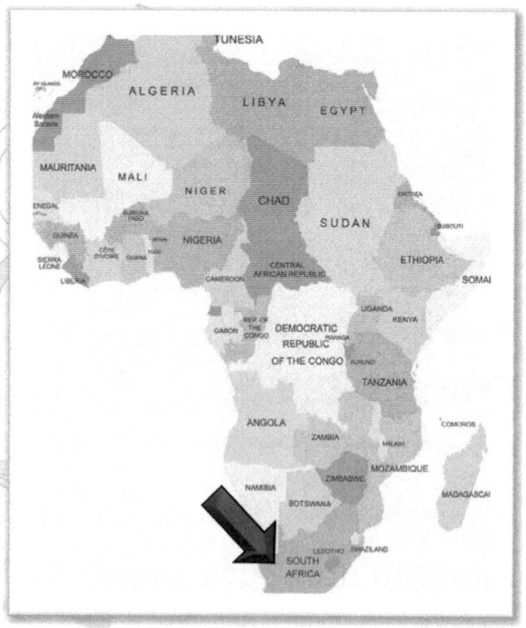

World Zinc Mine Production Capacity Trend by Current Mine Status 2018 - 2021

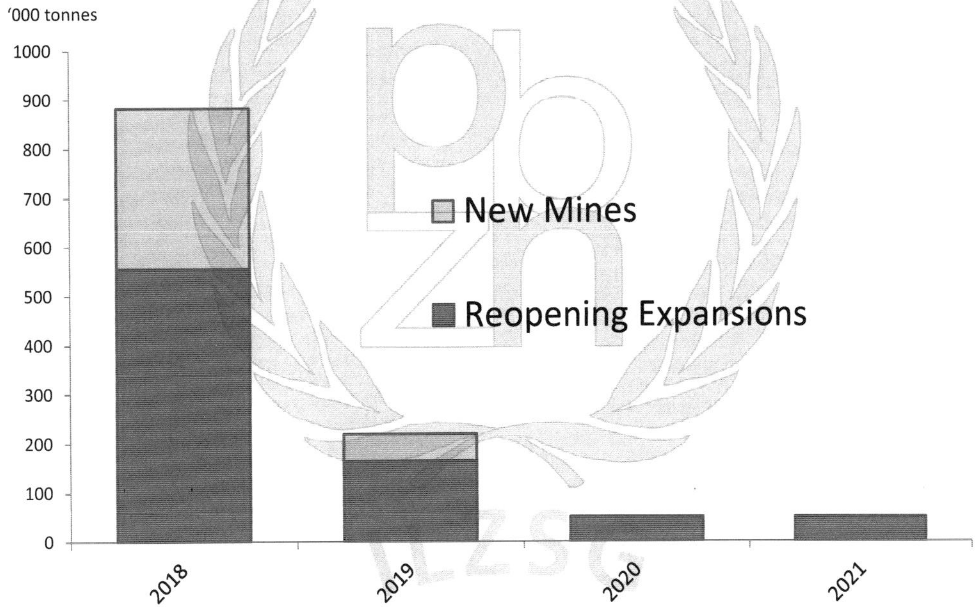

www.ilzsg.org International Lead and Zinc Study Group Source: ILZSG 25

New Committed Mine Capacity by Country 2018 – 2021

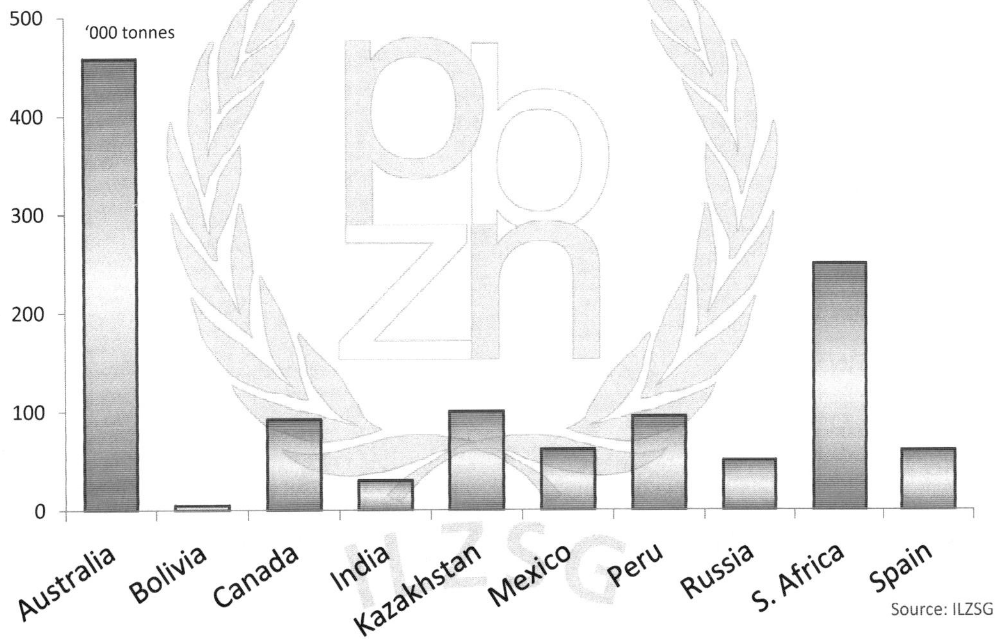

Source: ILZSG

www.ilzsg.org International Lead and Zinc Study Group 26

Presentation Outline

- ILZSG Overview
- Zinc Mine Supply General Trends
- Cutbacks and Closures
- Recent Openings and Committed Additions
- Longer Term Possibilties
- Conclusions

Selected Zinc Mine Projects Under Consideration

Mine	Estimated Annual Capacity (Zinc contained)	Possible Opening
Oued-Amizour (Tala Hamza), Algeria	164,000t	2022
Admiral Bay, Australia	450,000t	2024
Dairi, Indonesia	100,000t	2022
Izok Lake, Australia	80,000t	2021
San Gregorio, Peru	105,000t	2022
Mehdiabad, Iran	400,000t	2022
Citronen, Greenland	170,000t	2024
Pavlovskoye, Russia	110,000t	2021

Source: New Mines and Smelters 2017 and 2018 Reports, ILZSG

Mehdiabad Project, Iran

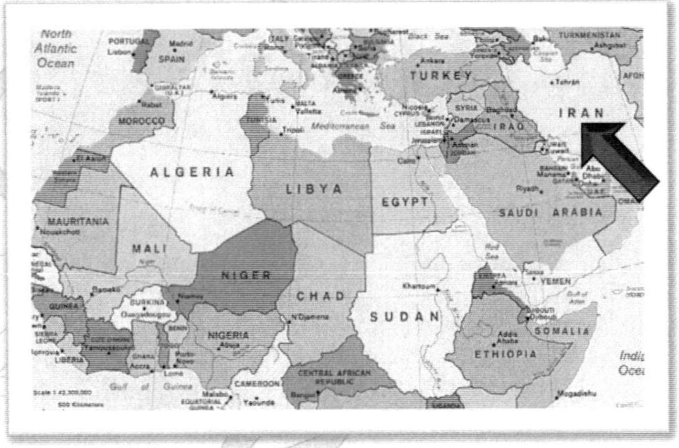

- World's largest undeveloped zinc deposit
- Possible 400kt/year zinc
- 25% of resource is oxide
- Private consortium led my Mobin Mining studying development. IMIDRO also have share.

Selected Zinc Mine Projects Under Consideration 2

Source: New Mines and Smelters 2017 and 2018 Reports, ILZSG

Mine	Estimated Annual Capacity (Zinc contained)	Possible Opening
Zhairem, Kazakhstan	160,000t	2020
Mount Isa Super Pit, Australia	300,000t	2022/23
Asmara, Eritrea	85,000t	2019/20
Pallas Green, Ireland	150,000t	2022/23
Ozernoye, Russian Federation	440,000t	2022
Khnaiguiyah, Saudi Arabia	70,000t	2022
Kipushi, DR Congo	225,000t	2020/21
Hermoza, United States	100,000t	2021

Ozernoye Project, Russia

- Located in Rep. of Buryatia in Southern Siberia
- Framework agreement between NFC & East Siberian Metals agreed in 2012
- Possible 440Kt/year zinc by 2022
- Total cost US$ 1.3 billion

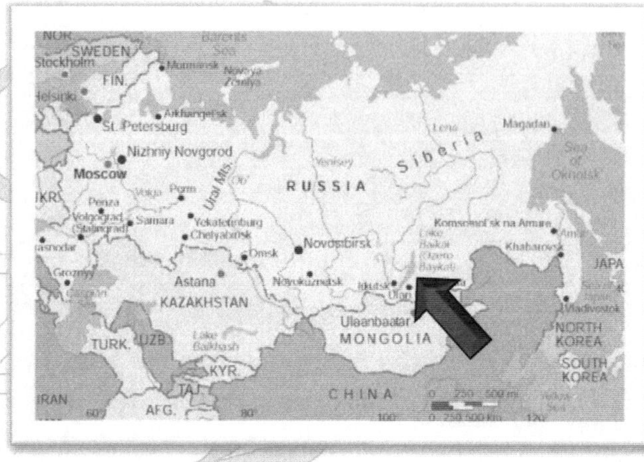

Selected Zn Mine Projects

Capacity > 100kt

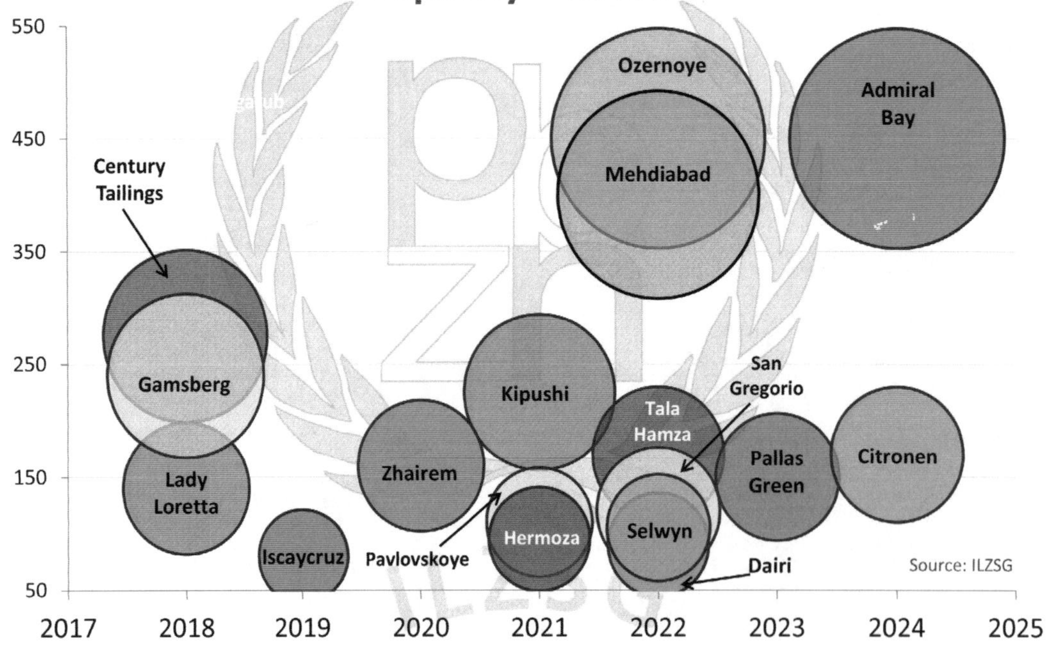

More Zinc will be Recovered from EAF Dust

Recently completed projects include:

- Porto Vesme, Italy
- American Zinc Recycling (*formerly Horsehead*), Mooresboro *(technical problems under review)*
- Korea Zinc (ZincOx Resources)
- Global Steel Dust, Saudi Arabia
- TSU, Taiwan

Presentation Outline

- ILZSG Overview
- Zinc Mine Supply General Trends
- Cutbacks and Closures
- Recent Openings and Committed Additions
- Longer Term Possibilties
- Conclusions

Conclusions

- Substantial reduction in ex-China zinc mine output in 2015 resulted in rapid tightening of zinc concs market
- 880kt of new capacity currently committed to open in 2018 of which 2/3 will be expansions or re-openings
- Only 319kt of new capacity so far committed over period 2019-2021
- 60% of all currently committed capacity located in Australia and South Africa
- A number of large projects under consideration from 2021 but questions remain over timing and viability for most

THANK YOU!

FIREWEEDZINC

Advancing the Macmillan Pass Zinc Project: Challenges and Opportunities

February 2018 TSX-V:FWZ

Cautionary Statements

The following statements are required by Canadian securities legislation:

Forward-Looking Statements:

This presentation may contain certain "forward-looking" statements and information relating to Fireweed Zinc that are based on the beliefs of Fireweed Zinc management, as well as assumptions made by and information currently available to Fireweed Zinc management. Such statements reflect the current risks, uncertainties and assumptions related to certain factors including but not limited to, without limitations, exploration and development risks, expenditure and financing requirements, general economic conditions, changes in financial markets, the ability to properly and efficiently staff the Company's operations, the sufficiency of working capital and funding for continued operations, title matters, First Nations relations, operating hazards, political and economic factors, competitive factors, metal prices, relationships with vendors and strategic partners, governmental regulations and supervision, permitting, seasonality and weather, technological change, industry practices, and one-time events. Should any one or more risks or uncertainties materialize or change, or should any underlying assumptions prove incorrect, actual results and forward-looking statements may vary materially from those described herein. Fireweed does not assume the obligation to update any forward-looking statement or forward-looking information except as required by law.

NI43-101 Qualified Person:

Brandon Macdonald P.Geo ,CEO and Director of Fireweed Zinc, and a Qualified Person under the meaning of Canadian National Instrument 43-101, is responsible for the technical information in this presentation.

Overview

Fireweed Zinc Overview

Fireweed is a new company that completed an IPO in 2017 and is built around the Macmillan Pass Zinc-Lead-Silver project in Yukon Territory, Canada

2018 Macmillan Pass Resource Update

The January 2018 Mineral Resource Estimate puts Macmillan pass among the world's most significant zinc resources.

Huge Potential for Resource Expansion and New Discoveries

The Macmillan Pass Project already has not only world-class established mineralisation that is open for expansion, but also big exploration upside.

Getting to Production

Permitting, environmental monitoring, community relations and project economics all play a factor in getting a project to production.

Intro to Fireweed and Macmillan Pass

About Fireweed Zinc

Building a New Zinc Company

- Public Canadian Corporation headquartered in Vancouver
- Focused on the Macmillan Pass Project
- Owns 100% of Tom & Jason deposits with an additional option to acquire 100% of Mac claims
- T&J represent a substantial zinc-lead-silver resource with excellent exploration potential and moderate exploration costs
- Located in politically stable Canada, and the mining-friendly Yukon territory
- Well funded: recent $12.5M placement over-subscribed

Share Structure	
Issued & Outstanding	**30,046,776**
Agent's Warrants	639,000
Options	1,610,000
Performance Shares	4,000,000
Fully-Diluted	**36,295,776**

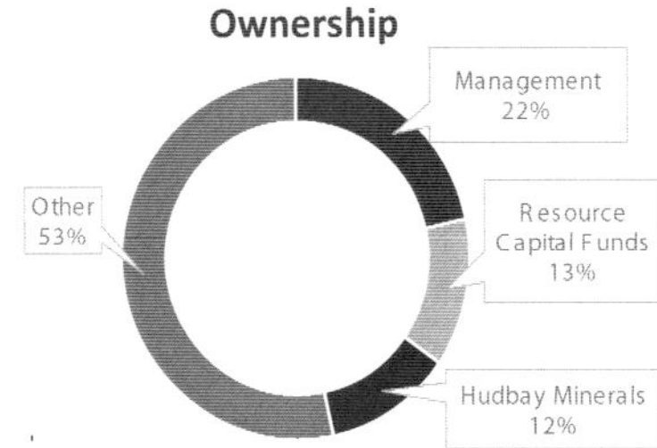

The Macmillan Pass Project

Building on a Substantial Zinc-Lead Resource

- Located in Yukon, Canada, mining friendly jurisdiction with road access via Canol Road and regional airstrip
- Contains two known SEDEX Deposits
- 1951 - Tom discovered by Hudbay
- 1974 - Jason found by Ogilvie JV
- 2007 - Hudbay acquires Jason, uniting the two properties
- 2016 – Fireweed Signs Option on Tom & Jason Properties
- 2018 – Fireweed exercises option to acquire 100% of T&J
- ~75,000m of drilling on the two properties since initial discovery
- Combined Resource of
  - 11.21 Mt Indicated @ 9.61% ZnEq
    - 6.59% Zn, 2.48% Pb, 21.33 g/t Ag
  - 39.47 Mt Inferred @ 10.00% ZnEq
    - 5.84% Zn, 3.14% Pb, 38.15 g/t Ag

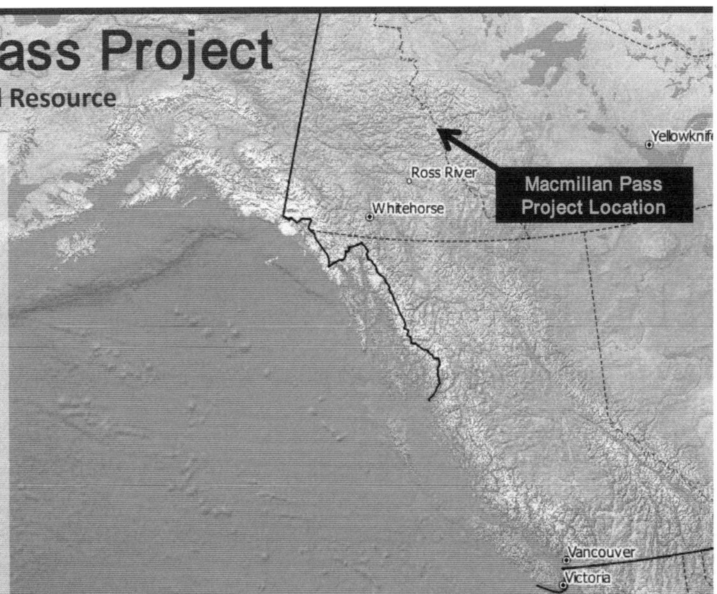

Management Team

Industry Veterans with Diverse Experience

John Robins
Executive Chairman & Director

- Founder, Executive Chairman & Director of Kaminak Gold Corporation
  - $520M T/O by Goldcorp
- Current director of Northern Empire Resources Corp., K2 Gold Corporation, Bluestone Resources Inc.
  - In 2017 Mr. Robin's companies raised >$100M
- Winner of AMEBC's H.H. "Spud" Huestis Award 2008

Brandon Macdonald
CEO & Director

- Current director of NorthIsle Copper & Gold Inc, Commander Resources Ltd
- Ex Macquarie Bank
- BSc Geology UBC, MBA Oxford University
- Long history of work in Yukon including zinc projects
- Originally hails from Ross River, Yukon, closest community to Tom & Jason

George Gorzynski
Director
- Director, VPX IMPACT Silver

Richard Hajdukiewicz
Director
- Ex Goldman & Deutsche

Dan Rogness
Director
- Ex Teck

Adrian Rothwell
Director
- CEO Kore Mining

FIREWEEDZINC 7 TSX-V:FWZ

Yukon Territory

150+ Year History of Mining

- Historical Mining Jurisdiction
  - Active mining from Klondike Gold Rush until now
- Geopolitically Stable & Safe
  - Federal & Territorial Government, alongside industry, committed to nearly $500M for resource roads
  - YESAB: 1-stop-shop for permitting
  - 10+ Mining Licenses granted in last 5 years
- Progressive Partnerships:
  - Industry
  - First Nations
  - Government
- Highly Prospective and Underexplored
  - "World's Greatest Geological Potential"
  - Only 12% of Yukon's land mass has been explored
  - World-Class geological survey with commitment to research and public data

FIREWEEDZINC 8 TSX-V:FWZ

2018 Macmillan Pass Resource Update

FIREWEEDZINC 9 TSX-V:FWZ

World-Class Zinc-Lead-Silver Resource

One of the Largest Undeveloped Zinc Resources in the World

Brand New 2018 Resource Update

	tonnes	Zn %	Pb %	Ag g/t	ZnEq%	Contained	Zn Mt	Pb Mt	Ag MOz
Tom	8,713,000	6.12%	2.68%	26.80	9.49%				
Jason	2,494,000	8.25%	1.76%	2.22	10.04%				
Indicated Total	**11,207,000**	**6.59%**	**2.48%**	**21.33**	**9.61%**		0.74	0.28	7.69
Tom	23,222,000	6.27%	2.96%	36.49	10.19%				
Jason	16,243,000	5.23%	3.39%	40.53	9.72%				
Inferred Total	**39,4645,000**	**5.84%**	**3.14%**	**38.15**	**10.00%**		2.30	1.24	48.41

- Now one of the largest zinc deposits in the world held by a junior
- Not just compelling size, but compelling average grade

FIREWEEDZINC 10 TSX-V:FWZ

World-Class Zinc-Lead-Silver Resource

Substantial tonnage at a variety of cut-offs

NSR $CAD	Indicated tonnes	Zn	Pb	Ag	Zn Eq
45	11,432,146	6.52%	2.44%	20.96	9.49%
65	**11,207,004**	**6.59%**	**2.48%**	**21.33**	**9.61%**
85	10,303,841	6.81%	2.65%	22.92	10.04%
105	7,625,412	7.49%	3.22%	29.77	11.48%

NSR $CAD	Inferred tonnes	Zn	Pb	Ag	Zn Eq
45	43,143,733	5.55%	2.95%	35.45	9.44%
65	**39,464,855**	**5.84%**	**3.14%**	**38.15**	**10.00%**
85	33,176,701	6.26%	3.56%	43.99	11.01%
105	24,478,058	6.82%	4.50%	56.34	12.83%

FIREWEEDZINC 11 TSX-V:FWZ

New Resource Model: Tom Deposit

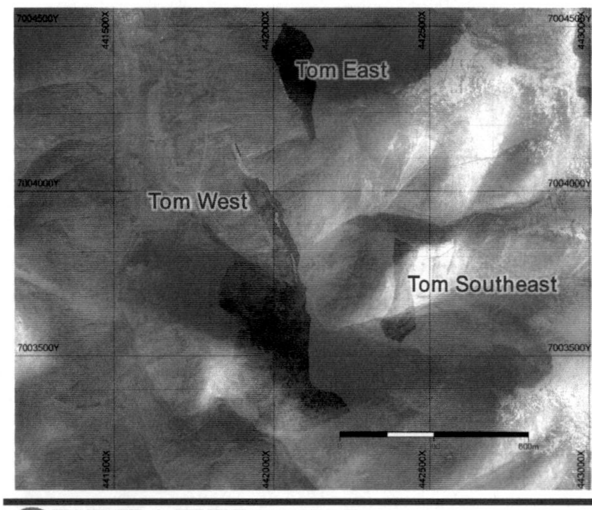

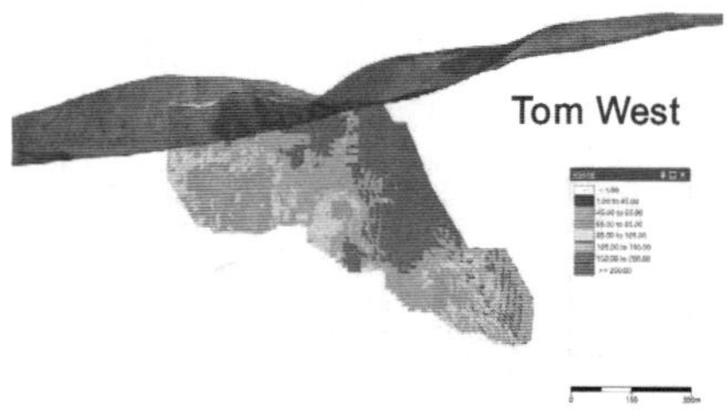

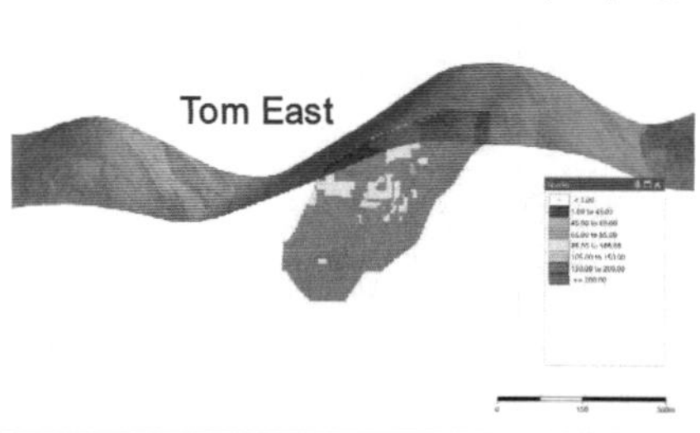

FIREWEEDZINC 12 TSX-V:FWZ

New Resource Model: Jason Deposit

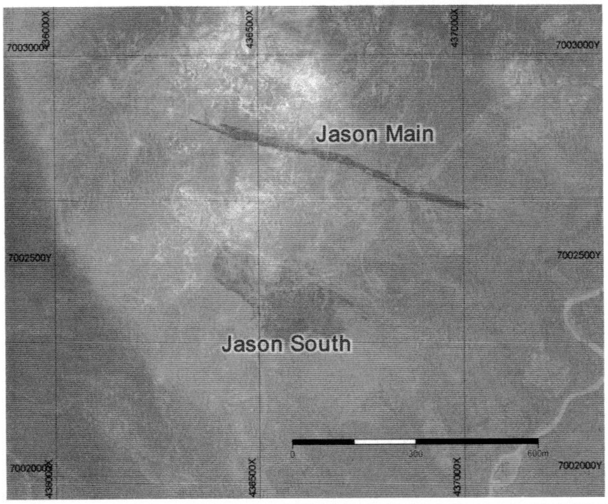

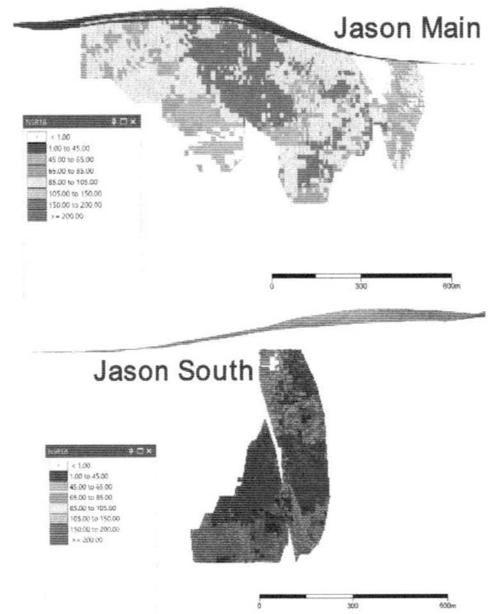

FIREWEEDZINC 13 TSX-V:FWZ

Macmillan Pass: Already Big and Getting Bigger

FIREWEEDZINC 14 TSX-V:FWZ

Nearly 70 Years of Data

66 years of exploration data:

Surface mapping, underground mapping, surficial mapping, trenching, prospecting

Geophysics: gravity, magnetics, EM (Max-min, Crone EM, VLF-EM, DigHEM, HLEM), CSAMT, IP,

Geochemistry: soils, stream sediments, water sampling, MMI, biogeochem

Prior to 2017:

- Tom: 33,495 m between 212 drill holes
- Jason 37,925 m between 128 drill holes

Underground lateral and decline development at Tom (3,645 m total)

Bulk sampling, metallurgy, scoping studies, feasibility studies, resource estimates

Ph.D., M.Sc. and B.Sc. Theses

Scientific publications

Government publications

Air photos, fossil studies, water studies, environmental studies, wildlife studies, engineering studies, archaeological studies

FIREWEEDZINC 15 TSX-V:FWZ

Low-Hanging Fruit: Mineralization Expansion

Tom West Deposit

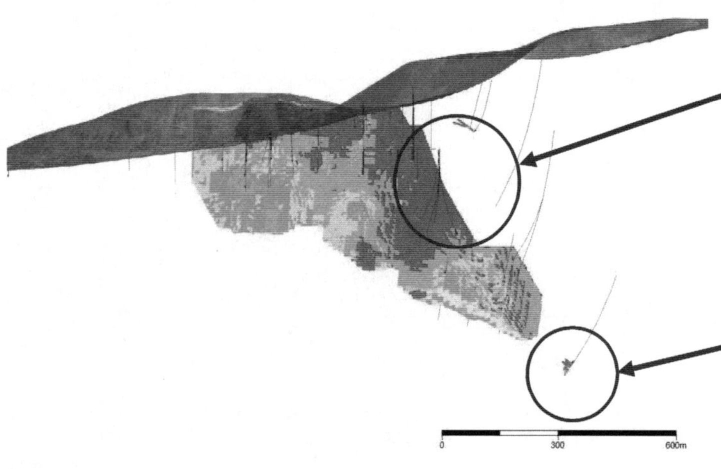

Excellent Upside Potential

- Significant potential to expand the known high-grade mineralization to the south
- Very little deep drilling has been done
  - Most zones extend to depth and have not been tested
- Potential for expansion along trend
  - Big step-out in 1990 yielded 14.6m of 14.1% Zinc (no follow-up)

FIREWEEDZINC 16 TSX-V:FWZ

Low-Hanging Fruit: Mineralization Expansion

Jason Syncline

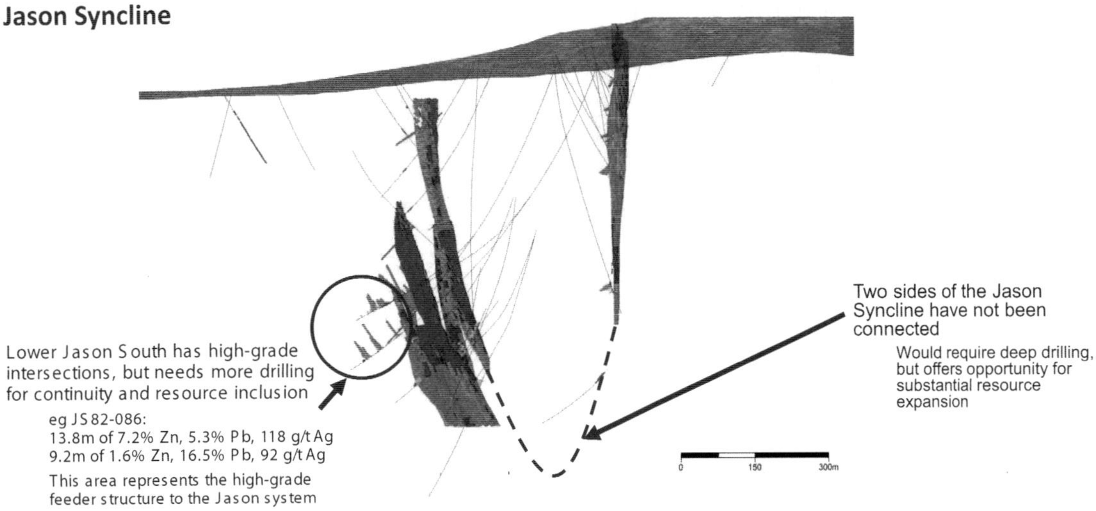

FIREWEEDZINC 17 TSX-V:FWZ

Big Land Position for Discovery

Exploration Potential over Large 220 Sq Km Land Position

- Together the Tom, Jason and Mac claims cover a large area of 21,939 Ha (53,920 acres)
- Footprint of the known deposits is very small compared to property size
- Much of the rest of the ground remains little explored using modern exploration techniques
  - Most recent large-scale field exploration programs at Tom and Jason were more than 25 years ago
- Satellite zones represent immediate exploration upside
  - End Zone: Hole JS80-59 intersected 3.6% Zn, 10.2% Pb and 83.2g/t Ag over 19.4m
  - End Zone: Fireweed is first company to own both sides of this zone which extends over property boundary
  - End Zone: a nascent 3rd SEDEX system on the property? Needs exploration to test potential.

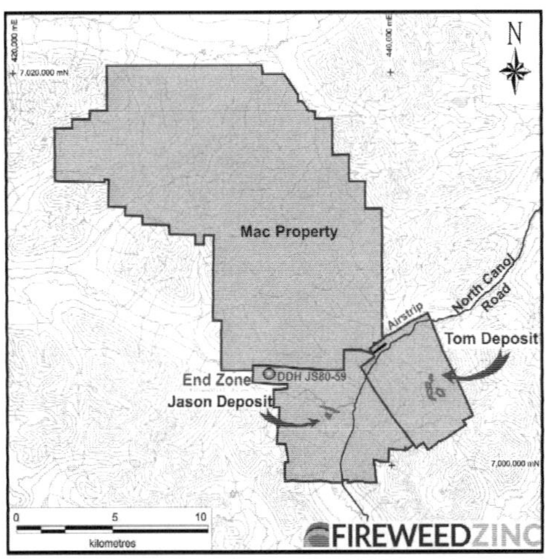

FIREWEEDZINC 18 TSX-V:FWZ

World-Class Exploration Potential

Ample Potential for New Discoveries

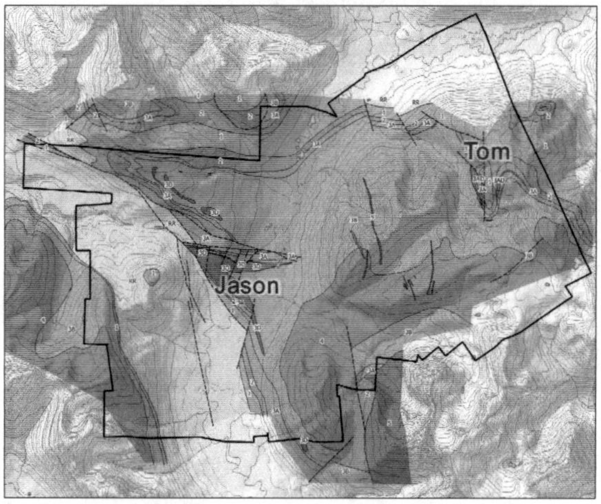

- Historic work has largely been focused on proving up known mineralization
  - Previous exploration programs left untested exploration targets
- Exploration vectors point to new areas of property
  - Geochemistry (stream sediments, water, soils)
  - Stratigraphy (basin edge)
- Large land package provides ample room for new discoveries
  - Modern technology allows exploration in areas ignored before
- Three known productive systems in project area, how many more?

" SEDEX" Exhalative Model

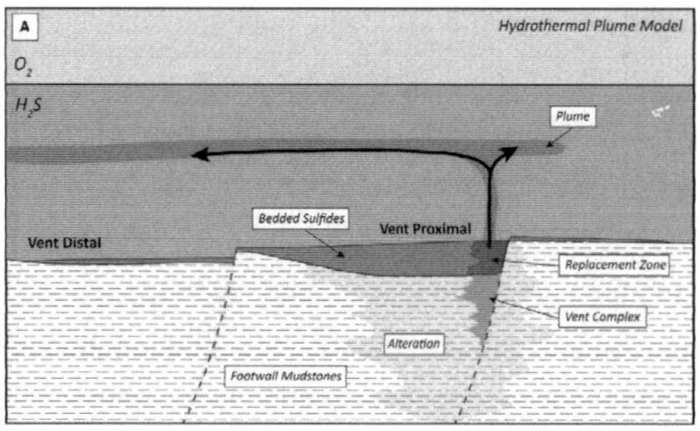

- "Old School"
- Originally coined at Tom
- Euxinic bottom waters are key
- Targets distinct stratigraphic interval, and euxinic sub-basins

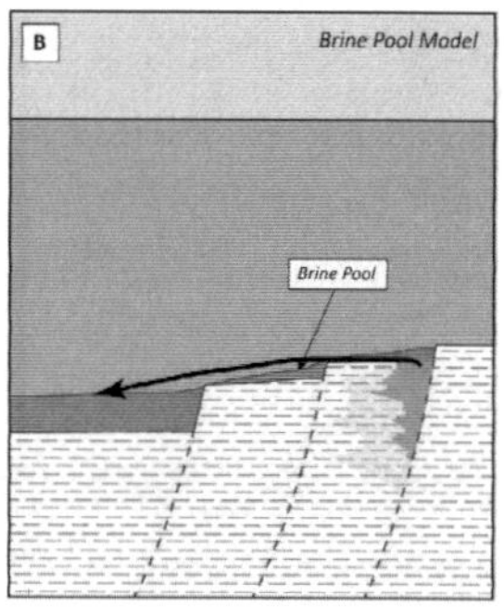

(from Magnall 2015)

Replacement Model

- New school
- Barite and pyrite were replaced by Zn-Pb sulfides in the subsurface during diagenesis
- No need for euxinia
- Targets areas of barite/pyrite enrichment, possibly in older stratigraphy

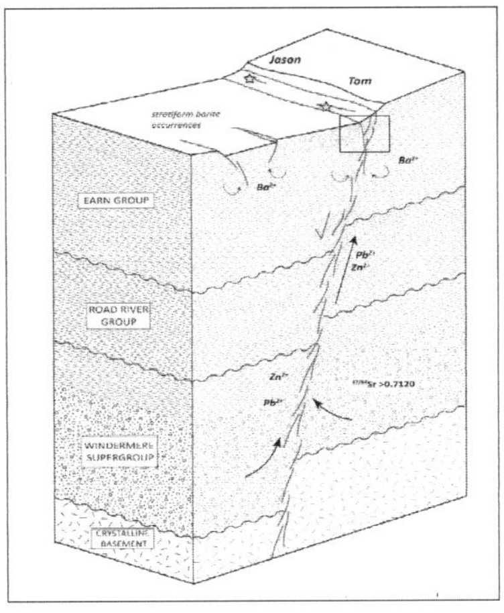

(from Magnall 2015)

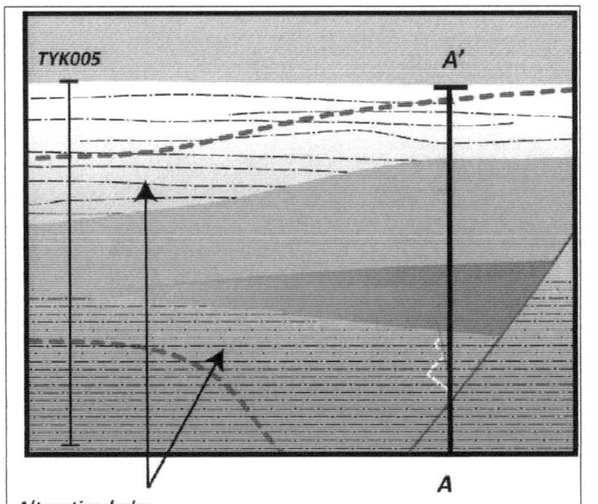

Alteration halo:
$K_2O/Al_2O_3 < 0.28$; Ba = 1000ppm to 7wt %; increasing $\delta^{34}S$ values

(from Magnall 2015)

FIREWEEDZINC TSX-V:FWZ

Extent of Ground Cover

- Much of the property is covered.
- Thickness variations are likely rapid.
- This may easily mask geophysical and geochemical anomalies.
- Exploration in these areas has likely been overlooked in the past and therefore, is where the biggest opportunity lies for discovery.

Purple shows approximate covered areas

FIREWEEDZINC TSX-V:FWZ

VTEM : Electromagnetics (EM)

1D conductivity model: 50 m depth slice

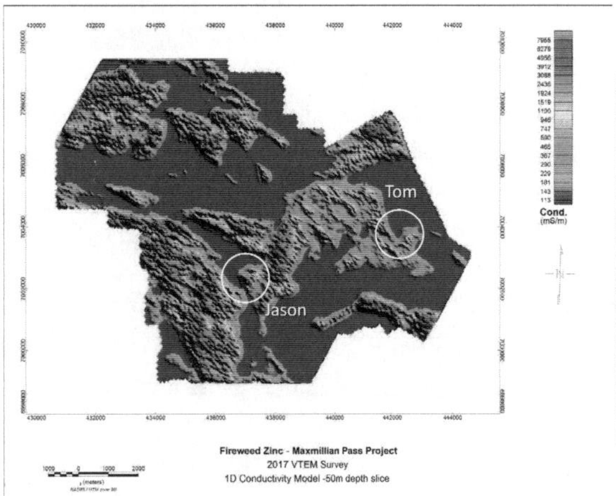

Landsat7 7-4-3 False Colour Composite Satellite Image with deposit outlines

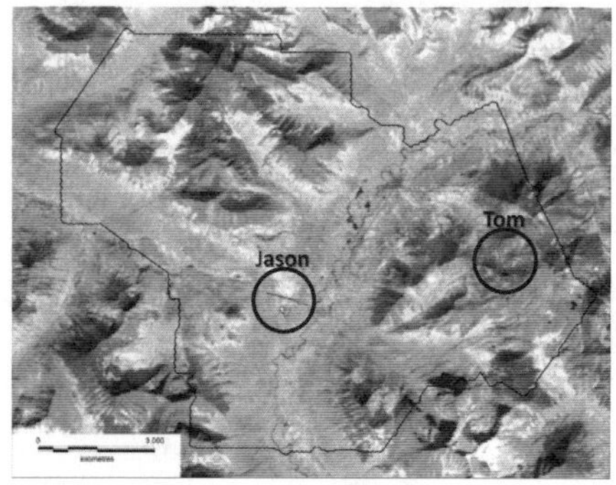

Stratigraphic units and structural breaks mapped well with the conductivity model developed from the VTEM data.

FIREWEED ZINC 23 TSX-V:FWZ

Geochemistry in Covered Areas

- Cover includes Quaternary and younger units in stratigraphy
- Lots of potential to try a new approach.
- Soil pH to detect reduced chimney above buried sulphides
  - Presence of calcrete at End Zone supports this.
- pXRF and soil pH give IN FIELD results for fast follow-up ahead of drilling
- Biogeochemistry?
- Soil gas hydrocarbons
- Highly targeted surveys as follow ups to EM-Gravity anomalies

FIREWEED ZINC 24 TSX-V:FWZ

Getting to Production

FIREWEEDZINC 25 TSX-V:FWZ

Infrastructure

Excellent Access for Exploration & Development

- Project is on the government maintained Canol Road, ~200 km from the town of Ross River
- Macmillan Pass airstrip within property boundary
- Infrastructure upgrades required to transport concentrate and operate a mine

FIREWEEDZINC 26 TSX-V:FWZ

Environment and CSR

Proactive Project Stewardship

- Environmental baseline studies began at site in 2000 and were expanded in 2008
- 12 surface water monitoring stations, 3 ground water wells, and 2 weather stations
- Nearest community is Ross River, population 350, home to Ross River Dena First Nations
- Relationship with Ross River Dena Council First Nations has been good with Fireweed engaging numerous FN-owned businesses in 2017 with expanded engagement planned in 2018
- CEO Brandon Macdonald is originally from Ross River and is familiar with the community

Metallurgy

Encouraging Results from Initial Study

- Preliminary Metallurgical Study Conducted in 2012
- Whole-core HQ from 4 holes in 2011, totalling 764 kg from the Tom West deposit
- Recoveries:
  - 83% of the Pb and 73% of the Ag in the Pb concentrate
  - 79% of the Zn and 12% of the Ag in the Zn concentrate
- Low iron in concentrates: 0.94% Fe in zinc concentrate, 2.29% Fe in lead concentrate
- Initial Concentrate results summarized below:

Product	Weight %	Assay %		
		Zn	Pb	Ag
Flotation Feed	100	5.15	3.18	34
Zinc Concentrate	7.0	58.5	1.76	56
Lead Concentrate	3.7	4.23	70.9	673

- Additional Metallurgical testing underway with results due soon

Near-Term Corporate Objectives

Focus on Advancing Project and Resource Expansion

Timeframe	Objective
2018 Q1 (Complete)	• **Updated 43-101 Resource**
2018 Q2	• **Preliminary Economic Assessment**
2018 Summer	• Commence large-scale drilling program with three goals: • Upgrade priority zones to M&I for Pre-Feasibility • Expand known zones through step-outs • Drill new targets and prove up entirely new deposits • Extensive regional exploration including Geochemistry, Geophysics,etc • Expand baseline environmental to kick off permitting process

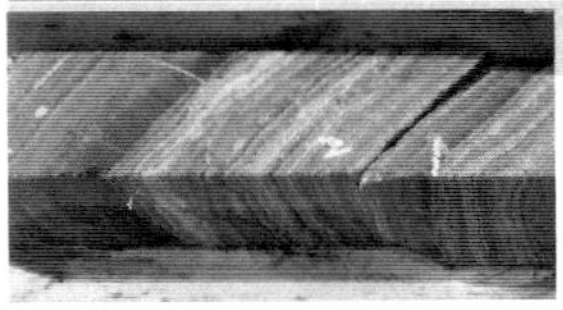

FIREWEEDZINC 29 TSX-V:FWZ

Photos

FIREWEEDZINC 30 TSX-V:FWZ

FIREWEEDZINC 31 TSX-V:FWZ

FIREWEEDZINC 32 TSX-V:FWZ

FIREWEEDZINC 33 TSX-V:FWZ

For More Information, Please Contact:

Brandon Macdonald
CEO & Director

Email: brandon@ fireweedzinc.com
Phone: +1 604 646 8360

Suite 1020 - 800 West Pender St.
Vancouver, BC V6C 2V6

FIREWEEDZINC 34 TSX-V:FWZ

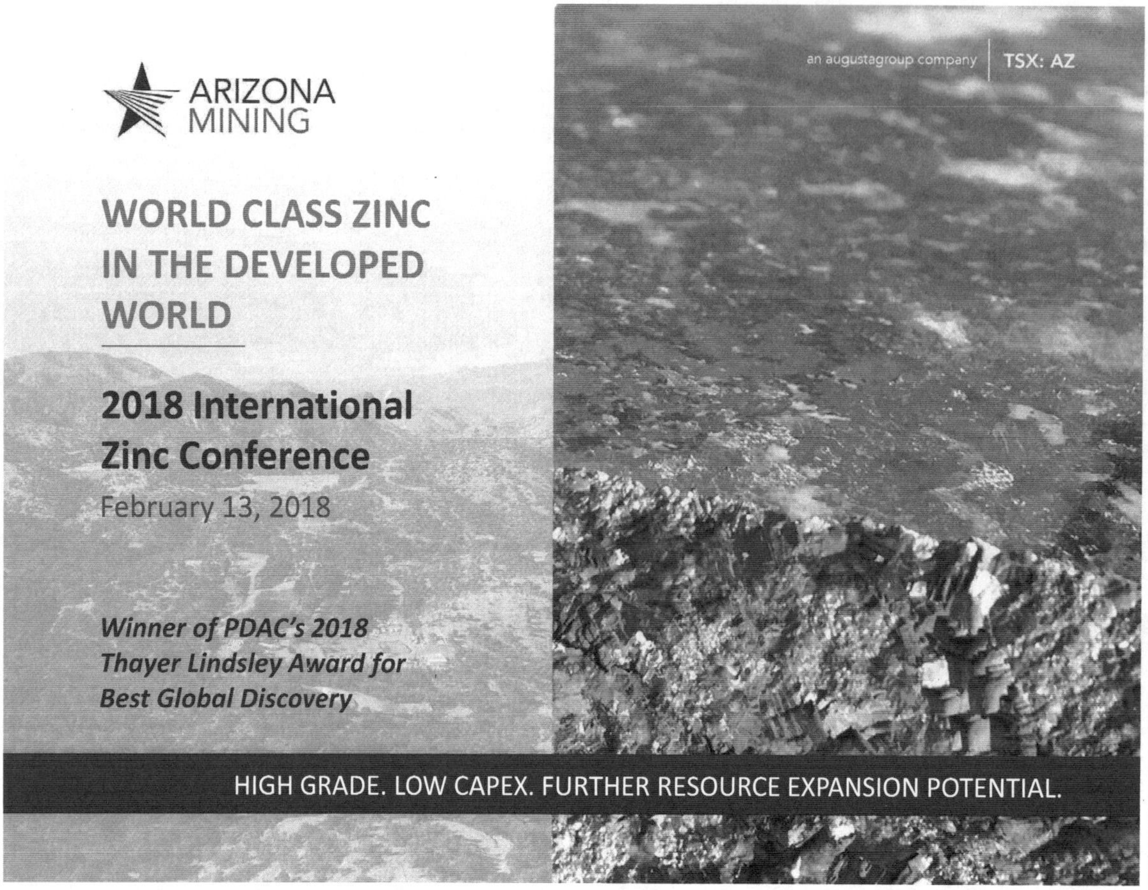

CAUTIONARY STATEMENT ON FORWARD-LOOKING INFORMATION

ARIZONA MINING

Certain information contained in this presentation constitutes forward-looking statements. All statements, other than statements of historical facts, are forward looking statements including statements with respect to the Company's intentions for its Hermosa Project in Arizona, USA including, without limitation, financing, future drilling and other work on the Taylor Deposit. The Company would also like to caution the reader that the preliminary economic assessment ("PEA") on the Company's Taylor Deposit that supports the technical feasibility or economic viability of the Taylor Deposit, including the marketability of the concentrate, mining methods, costs, recoveries and any other technical aspects related to the Taylor Deposit, is preliminary in nature and there is no certainty that the PEA will be realized. Forward-looking statements are often, but not always, identified by the use of words such as may, will, seek, anticipate, believe, plan, estimate, budget, schedule, forecast, project, expect, intend, or similar expressions.

The forward-looking statements are based on a number of assumptions which, while considered reasonable by Arizona Mining, are subject to risks and uncertainties. In addition to the assumptions herein, these assumptions include the assumptions described in Arizona Mining's management's discussion and analysis for the year ended December 31, 2016 ("MD&A"). Arizona Mining cautions readers that forward-looking statements involve and are subject to known and unknown risks, uncertainties and other factors which may cause actual results, performance or achievements to differ materially from those expressed in or implied by such forward-looking statements and forward- looking statements are not guarantees of future results, performance or achievement. These risks, uncertainties and factors include general business, economic, competitive, political, regulatory and social uncertainties; actual results of exploration activities and economic evaluations; fluctuations in currency exchange rates; changes in project parameters; changes in costs, including labour, infrastructure, operating and production costs; future prices of zinc, lead, silver and other minerals; variations of mineral grade or recovery rates; operating or technical difficulties in connection with exploration, development or mining activities, including the failure of plant, equipment or processes to operate as anticipated; delays in completion of exploration, development or construction activities; changes in government legislation and regulation; the ability to personnel and contractors; the speculative nature of, and the risks involved in, the exploration, development and mining business; and the factors discussed in the section entitled "Risks and Uncertainties" in the MD&A.

Although Arizona Mining has attempted to identify important risks, uncertainties and other factors that could cause actual performance, achievements, actions, events, results or conditions to differ materially from those expressed in or implied by the forward- looking information, there may be other risks, uncertainties and other factors that cause performance, achievements, actions, events, results or conditions to differ from those anticipated, estimated or intended. Unless otherwise indicated, forward-looking statements contained herein are as of the date hereof and Arizona Mining disclaims any obligation to update any forward-looking statements, whether as a result of new information, future events or results or otherwise, except as required by applicable law.

ABOUT RESERVES AND RESOURCES

This presentation uses the terms measured, indicated and inferred resources as a relative measure of the level of confidence in the resource estimate. Readers are cautioned that: (a) mineral resources are not economic mineral reserves; (b) the economic viability of resources that are not mineral reserves has not been demonstrated; and (c) it should not be assumed that further work on the stated resources will lead to mineral reserves that can be mined economically. In addition, inferred resources are considered too geologically speculative to have any economic considerations applied to them. It cannot be assumed that all or any part of an inferred mineral resource will ever be upgraded to a higher category. Under Canadian rules, estimates of inferred mineral resources may not form the basis of feasibility or pre- feasibility studies or economic studies except for certain preliminary economic assessments.

All dollars are United States Dollars except where noted

HIGH GRADE. LOW CAPEX. FURTHER RESOURCE EXPANSION POTENTIAL

ARIZONA MINING

- Taylor is the world's best zinc development project – world-class and growing high grade resource with district potential in Arizona
- Updated PEA shows US$2.0B NPV and rapid 1.6 year payback of pre-production capex
- 39% increase in M&I resources at similar grades – 29 year mine life
- Average annual production of 946 M lbs of zinc equivalent and 9.5 M oz of silver production in concentrate in first 5 years
- Key state permits received – an important milestone to execute mine development strategy
- Rapid development to production in 2020 at first quartile costs

HERMOSA PROJECT MINERAL RESOURCES

ARIZONA MINING

Taylor Deposit PEA Resources[1,2,3]

Classification	Million Short Tons	Zn (%)	Pb (%)	Ag (opt)	ZnEq (%)
Measured	15.2	4.0	4.0	1.6	9.6
Indicated	85.8	4.2	4.3	2.2	10.5
M&I	101.0	4.1	4.3	2.1	10.4
Inferred	43.6	3.9	4.8	3.4	11.9

Central Deposit PEA Resources

Classification	Million Short Tons	Zn (%)	Ag (opt)	Mn (%)	Oxval($/ton)
Measured	21.8	1.9	3.3	9.2	262
Indicated	41.7	2.3	1.7	9.8	257
M&I	63.5	2.2	2.3	9.6	259
Inferred	1.8	2.6	1.6	7.4	207

1. For a full list of drill holes included in the updated resource, please refer to www.arizonamining.com/projects/taylor-deposit/resources.
2. Mineral Resources are reported as of 30 November 2017. Stated at a cut-off grade of 4% ZnEq, and based on long-term prices, recovery and costs as follows: Prices of $1.00/lb for zinc, $0.95/lb for lead and $20.00/oz for silver; average processing recovery factors of 92% for zinc, 95% for lead, and 90% for silver; total operating costs are estimated to be $51/ton.
3. Mineral Resources are not Mineral Reserves and do not have demonstrated economic viability. There is no certainty that all or any part of mineral resources will be converted to Mineral Reserves. Inferred Mineral Resources are based on limited drilling which suggests the greatest uncertainty for a resource estimate and that geological continuity is only implied. Additional drilling will be required to verify geological and mineralization continuity and there is no certainty that all of the Inferred Resources will be converted to Measured and Indicated Resources. Quantity and grades are estimates and are rounded to reflect the fact that the resource estimate is an approximation.
Sources: January 1, 2018 technical report and AMC Mining Consultants

A DEVELOPED DISTRICT WITH MAJOR INFRASTRUCTURE

ARIZONA MINING

- Southern Arizona is a historic mining district with several large-scale base metal operations and projects
- Taylor is located close to major infrastructure, including power, gas, rail and Interstate Hwy 82

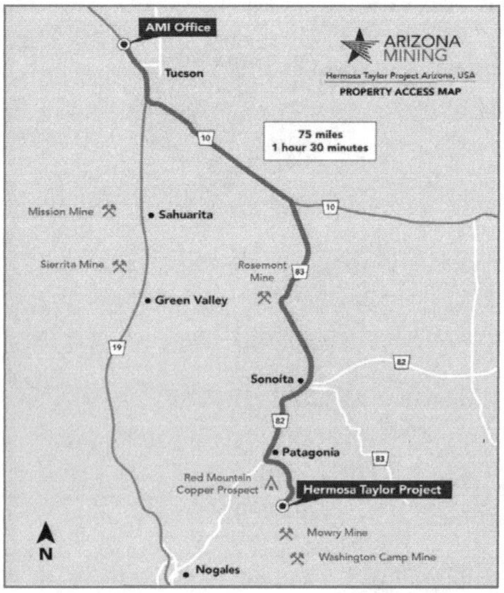

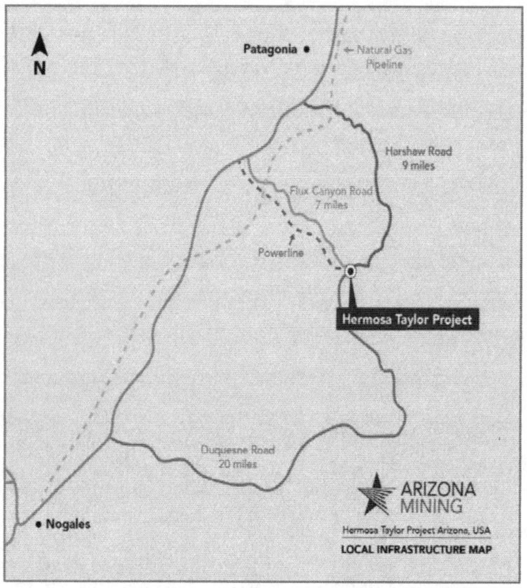

TWO DEPOSITS – DISTRICT SCALE EXPLORATION POTENTIAL

ARIZONA MINING

- The **Hermosa Project has 535 acres of private, patented mining claims** and **district exploration potential** within its 19,015 acres of unpatented mining claims
- **Taylor** zinc-lead-silver sulfide resource is the current focus
- Hermosa also contains the large **Central** silver-manganese zinc oxide resource which contains ~150 million ounces of contained silver[1]

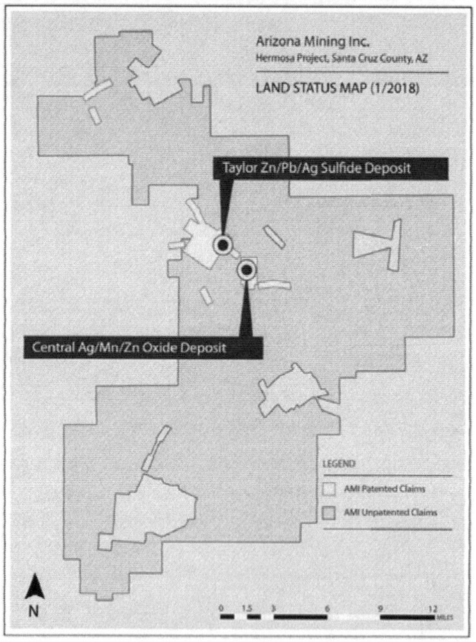

1. *January 1, 2018 technical report.*

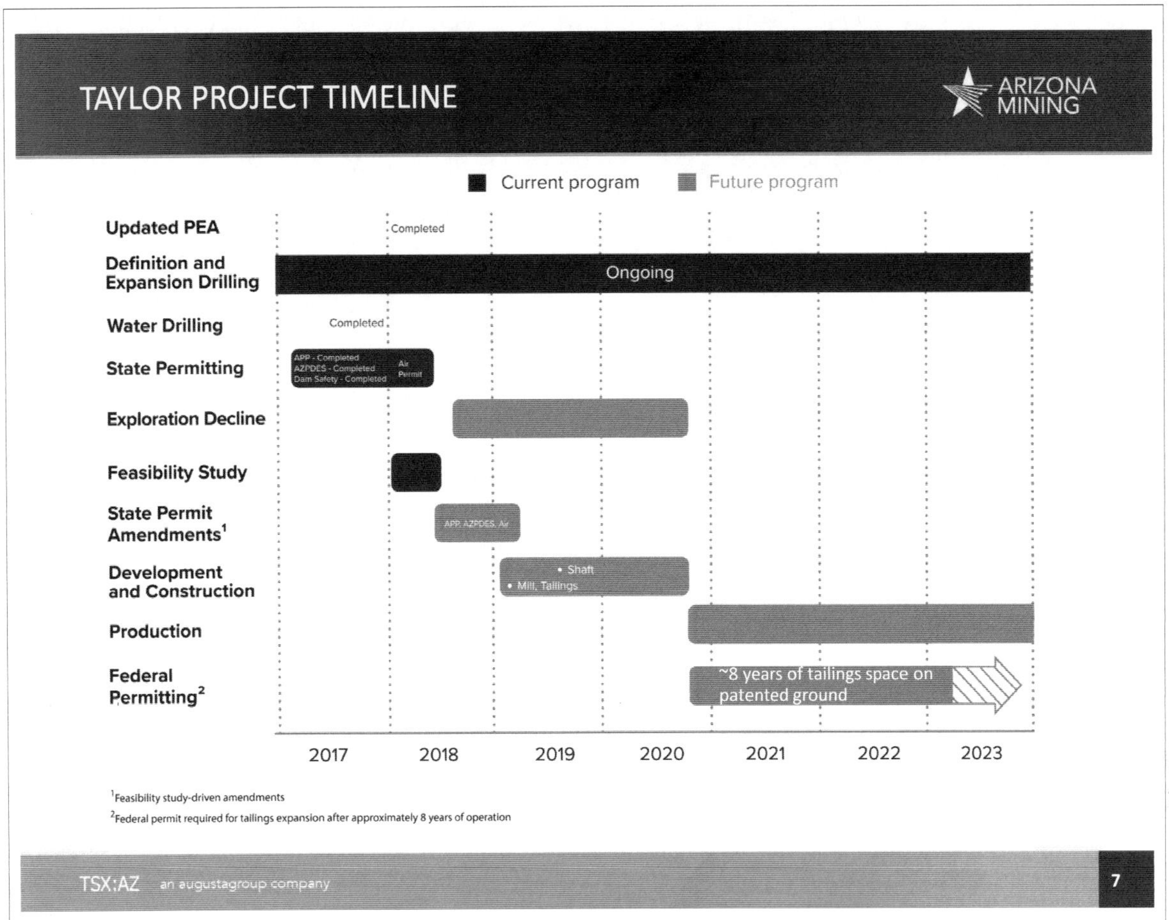
TAYLOR PROJECT TIMELINE
ARIZONA MINING
Current program
Future program
Updated PEA
Completed
Definition and Expansion Drilling
Ongoing
Water Drilling
Completed
State Permitting
APP - Completed
AZPDES - Completed
Dam Safety - Completed
Air Permit
Exploration Decline
Feasibility Study
State Permit Amendments[1]
APP, AZPDES, Air
Development and Construction
• Shaft
• Mill, Tailings
Production
Federal Permitting[2]
~8 years of tailings space on patented ground
2017
2018
2019
2020
2021
2022
2023
[1]Feasibility study-driven amendments
[2]Federal permit required for tailings expansion after approximately 8 years of operation
TSX:AZ
an augustagroup company
7

FUNDED TO EXECUTE OUR 2018 OBJECTIVES
ARIZONA MINING
❑ Obtain key state permits to begin construction of tailings facility and water treatment plant
❑ Obtain state Air Permit
❑ Begin exploration decline
❑ Test the potential of the deposit
❑ Deliver a Feasibility Study in mid-2018
Construction of new water treatment plant underway
TSX:AZ
an augustagroup company
APPENDIX
8

+$2B OF CUMULATIVE AFTER-TAX CASH FLOW IN 1ST 5 YEARS[1]

ARIZONA MINING

- Four payback periods in first five years
- Superior financial metrics should ensure an optimal mix of project funding

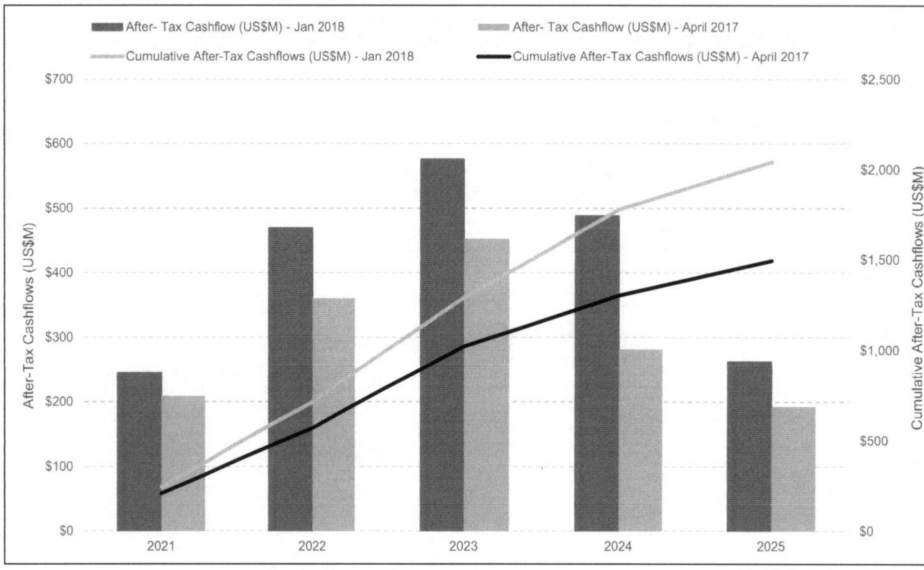

1. Cumulative after-tax cash flows begin after first production.
Sources: January 1, 2018 technical report and Arizona Mining Inc.

TAYLOR WILL BE A TOP 5 PRODUCER OF ZINC EQUIVALENT

ARIZONA MINING

- First five year average of 946 million pounds of zinc equivalent production

Top 15 Zinc Mines Globally Ranked by 2018 ZnEq Production

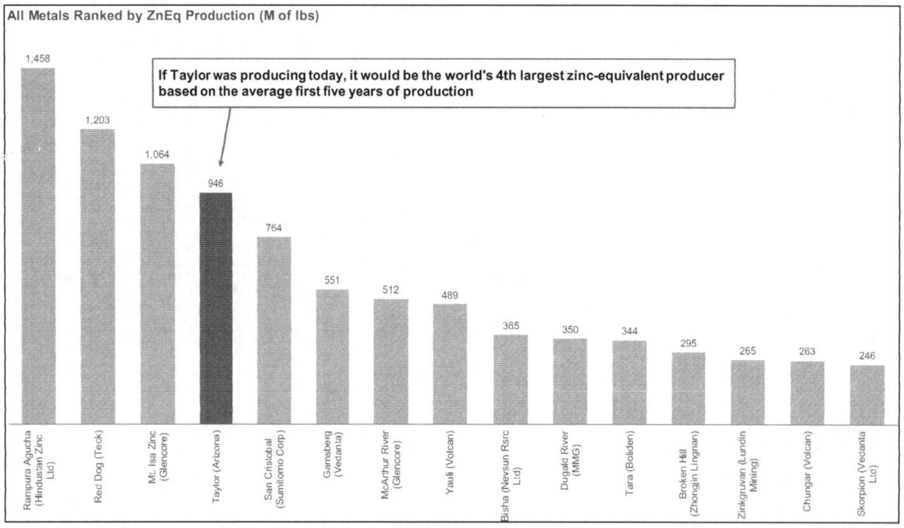

Notes:
1. Assumes metal prices of US$1.10/lb Zn, US$1.00/lb Pb, US$20.00/oz Ag, $1,250/oz Au.
2. ZnEq production estimates from SNL Metals & Mining for 2018 and illustrated for mines wherein>50% of the vaue of combined metal production is sourced from zinc.
Source: SNL Metals & Mining.

LOW CAPITAL INTENSITY

ARIZONA MINING

- 10,000 ton per day underground operation with decline and shaft
- Modest pre-production capex of $519 million incorporates accelerated spend for shaft development and additional equipment to access higher grade Taylor Deeps ore early in the mine plan
- First quartile C1 zinc co-product total cash costs of $0.49/lb[1]
- All-in sustaining ZnEq costs unchanged at $0.61/lb[1]
- Feasibility study by mid-2018

CAPEX ($US M)	PRE-PRODUCTION	SUSTAINING
Underground development/infrastructure	$58	$548
Mining equipment	$44	$95
Shaft	$126	$50
Process plant	$98	$8
Tailings facility	$19	$24
Site infrastructure	$71	-
EPCM, owners, other	$39	-
Contingency	$64	-
TOTAL	**$519**	**$725**

1. *Based on long term price assumptions of $1.10/lb zinc, $1.00/lb lead and $20/oz silver.*

HIGH GRADE CONCENTRATES

ARIZONA MINING

- Very coarse-grained ore with high recoveries

	JAN 2018
Recoveries in Zinc Concentrate	
Zinc	92.7
Silver	23.2
Recoveries In Lead Concentrate	
Lead	95.4
Silver	69.3
Concentrate Grades[1]	
Lead in Concentrate	69.7
Zinc in Concentrate	56.1

1. *Per short ton.*
2. *At a $1.00 per pound zinc price*

Source: SGS Lakefield

- PEA assumes a $13/dry metric tonne zinc treatment charge for average 1.35% manganese content or ~1% of zinc concentrate value of +$1,000/DMT[2]
- Updated PEA shows coarser grind size
- Further metallurgical testing is underway to test all stratigraphic horizons as part of the feasibility study

TAYLOR DEPOSIT – INDICATIVE SPECIFICATIONS

ARIZONA MINING

Metals/Minerals	Zn Concs	Pb Concs
Zn	55-59%	1-2.5%
Pb	1-2%	76-80%
Ag	50-300 gms/dmt	650-1500 gms/dmt
Cd	0.1-0.2%	0.01-0.02%
Mn	1-1.17%	0.2-0.3%
Cu	0.7-1.1%	0.2-0.5%
Fe	2-2.5%	0.6-3%
S	29-32%	12-14%
As	0.01-0.12%	<0.1%
Sb	0.05-0.2%	0.1-0.3%
Bi	<400 ppm	300-500 ppm
F	100-200 ppm	50-200 ppm
Hg	2-5 ppm	1-2 ppm
SiO2	2.5-3.5%	1.3-1.5%

TSX:AZ an augustagroup company 13

TAYLOR DEPOSIT – CONCENTRATE TONNAGE

ARIZONA MINING

Year	**Concentrate tonnage (DMT)**	
	Zinc	**Lead**
2021	270,000	171,000
2022	425,000	268,000
2023	525,000	361,000
2024	432,000	334,000
2025	271,000	219,000
First 5 Year Average	385,000	270,000
First 10 Year Average	316,000	226,000
LOM Average	221,000	198,000

TSX:AZ an augustagroup company 14

TAYLOR DEPOSIT – PAYABLE PRODUCTION

ARIZONA MINING

Year	Payable Production				
	Zinc		Lead		Silver
	M lbs	K tonnes	M lbs	K tonnes	M oz
2021	258	117	227	103	4.3
2022	405	184	355	161	7.5
2023	501	227	478	217	10.4
2024	412	187	442	200	9.7
2025	259	117	290	132	6.2
First 5 Year Average	367	166	358	162	7.6
First 10 Year Average	301	137	299	136	6.1
LOM Average	211	96	262	119	5.6

TSX:AZ an augustagroup company 15

TAYLOR – CONCENTRATE ROUTE

ARIZONA MINING

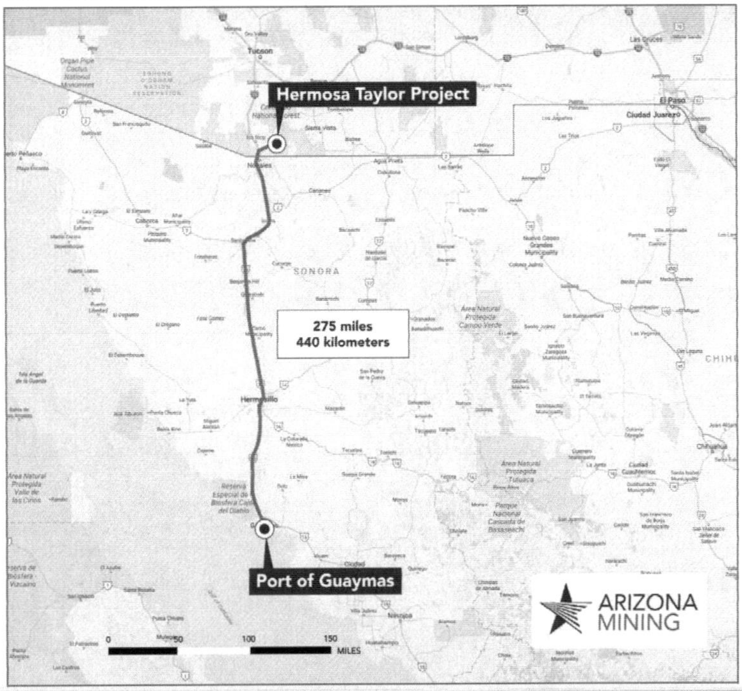

- Zinc and lead concentrates will likely be shipped to the deep-water port of Guaymas in Mexico
- Transportation costs are estimated at a total of $97 per dry metric tonne for both concentrates

TSX:AZ an augustagroup company APPENDIX 16

TAYLOR – GLOBAL SMELTERS

ARIZONA MINING

- Concentrates will then be shipped to various smelters around the world

2018 DRILLING IS FOCUSED ON HIGH PRIORITY TARGETS

ARIZONA MINING

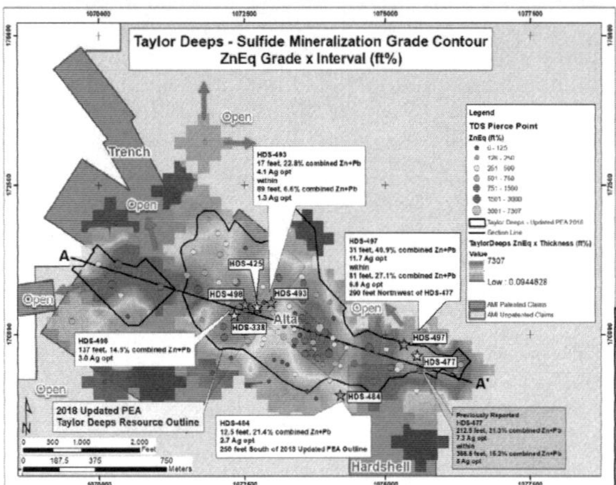

- Taylor deposit remains open in multiple directions
- Aggressive 2018 exploration program with 9 drill rigs - goal is to test the size and grade potential of Taylor Deeps and other high priority targets including on 21 acres of new private ground south of resource

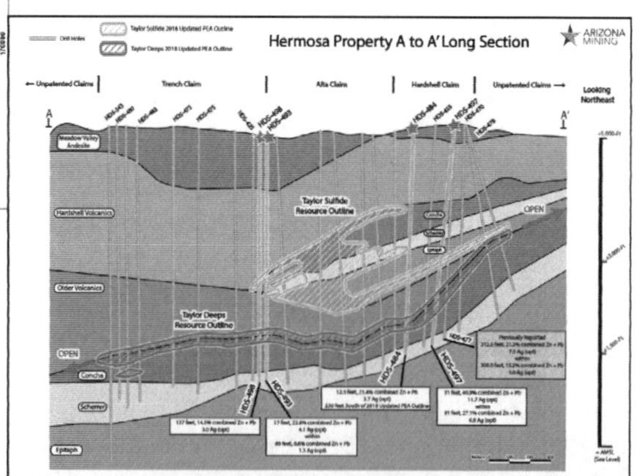

Meet the team: industry insiders disrupting base and precious metal concentrate trading

Boris Eykher
CEO & Co-Founder

Glencore
- Concentrate Trader
- CFO and Asset Manager Kazzinc

Pacific Corporate Group
- Private Equity Associate

International Council of Institutional Investors
- Director

Harvard Business School MBA
UC Berkeley BS in Electrical Engineering & Computer Science

Ilya Chernilovskiy
COO & Co-Founder

Glencore
- Asset Management
- Mining, metal production, commodity trading

Merrill Lynch
- Investment Banking Associate

AV Group
- Director of Strategy & Business Development

Harvard Business School MBA
Mendeleev University MSc in Chemical Technology

Diego Molina
Head of Business Development North & South America

Glencore (Switzerland, USA, Peru)
- Concentrate Trader – North & South America
- Contract execution for zinc and lead concentrates shipments
- Concentrate blending operations in Latin America

Lehigh University BA in International Relations

Max Weiland
Head of Freight

Torvald Klaveness, China
- Chartering Manager
- Freight trading and vessel chartering

Howe Robinson Partners – London and Shanghai
- Senior Shipbroker
- Servicing broad portfolio of clients across all dry vessel and market segments.

Warwick Business School – MBA with Distinction
East China University - Chinese Language and Literature

OPENMINERAL

2

Meet the team: industry insiders disrupting base and precious metal concentrate trading

Wen Shu
Head of Business Development - China

CITIC Metal
- Deputy General Manager of Base Metal department

Antofagasta Minerals
- Marketing Manager

LG International
- Concentrates Trader

Glencore
- Contract Execution for copper, zinc, lead concentrates

Beijing Technology and Business University
BA in Economics

David Gómez
Head of Operations Latin America

Glencore (Peru)
- Concentrate Trader – Peru, Bolivia, Mexico
- Contract admin for zinc, lead, copper, silver concentrates
- Concentrate blending operations in Latin America

Hochschild Mining (Peru)
- Senior analyst, silver concentrates and silver/gold dore

IE Business School MBA
Universidad del Pacífico BA in Business Administration

Gary Livshin
CTO

TimeTrade
- Senior Director of Development
- Building SaaS platform for Banking and Retail Industries

Talking Lights
- Technology & Innovation Team

20+ years experience in tech

Brandeis University
BA Computer Science

Ilona Kirienko
Head of Strategic Projects

Credit Suisse
- Marketing Head for PB Southern Europe

BBVA
- Strategy and Product Development Director for Small Business

McKinsey & Co
- Management consultant

20+ years experience in strategy and business development

Harvard Business School MBA
MGIMO (Moscow) MA in International Commerce

3

About Open Mineral Exchange (OME)

Open Mineral is a cloud-based exchange that makes it easy to competitively tender and source metal concentrates between mines and smelters.

OME ensures efficient trade execution and offers access to services supporting trade execution, including freight, surveying, insurance, and financing.

Key concentrate markets

Cu	Zn	Pb	Au	Ag
Copper	Zinc	Lead	Gold	Silver

- Exchange for direct and transparent trade of concentrates
- Success-based fee (fixed $/dmt)
- Based in Baar, Switzerland with offices in Beijing, Lima, and Moscow

OPENMINERAL

4

Concentrate market is inefficient and non-transparent

1. Trade of concentrates is dominated by commodity traders
2. Limited direct trade and communication between mines and smelters
3. Lack of market insights and useful analytics

So how do *you* make sure:

- That you get the best commercial terms?
- That you get the maximum profit on the commercial side?
- That there are no hidden costs?

5

Evolution of physical commodity trade

Open Mineral simplifies concentrate purchase and sale as well as execution by streamlining the processes and eliminating traders

6

Open Mineral Exchange benefits

Single platform for everything: tender, negotiate, execute the trade

100+ members

Benefits for Sellers & Buyers

1. Ensure you get the best market offer
2. Increase profit
3. Control trade execution and information flow

OPENMINERAL

7

Streamlining tenders and contracts execution

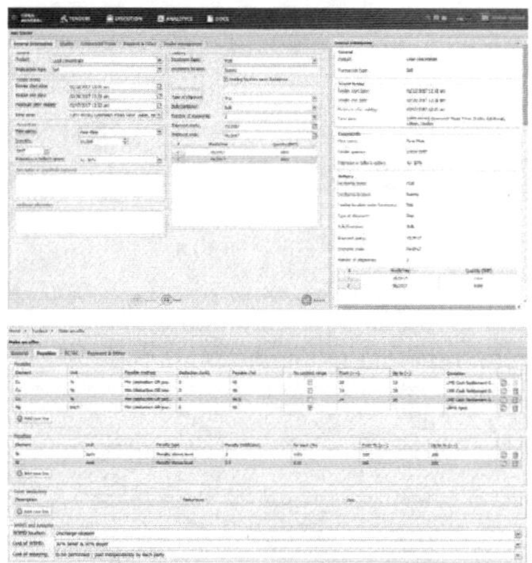

- Digitization of trade execution services
- Interactive calculator
- Execution reporting and analytics

8

Independent, Trusted, Private, Confidential

Trusted environment

- Thorough Know Your Customer (KYC) policy based on Thomson Reuters World Check for all registered members (mines, smelters, and service providers)
- Rejected ~60% of applicants
- Utilizing rating system to measure performance

Confidentiality & Privacy

- All user data is strictly confidential
- No disclosure of individual commercial data

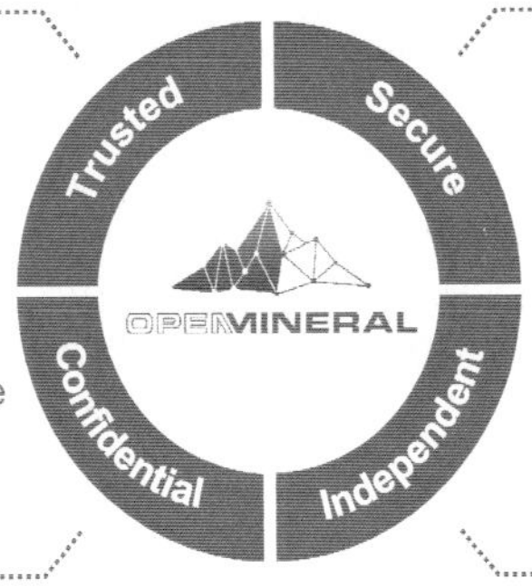

Security

- OME follows bank-level security standards
- Data is stored on Amazon Web Servers with best-in-class security
- Platform code is independently scanned for vulnerability by Veracode
- All transit data is SSL encrypted

Independent

- OME is fully independent and not affiliated with any commodity traders
- Built specifically for the needs of mines and smelters

9

Example – Mexican Pb/Ag Concentrate

Mexican Pb/Ag Concentrate			
Element	**Content**	**Unit**	**Spot Prices**
Pb	45	%	2300
Ag	10000	g/mt	17.80
Au	5	g/mt	1300
As	2	%	
Sb	1	%	
Bi	0	%	
Quantity (dmt)	**20000**		

OPENMINERAL

Paid to Miner: $/DMT

	Miner sells to Trader	Miner sells to Smelter	△	
Payable	**6,570**	**6,627**	**57**	
Lead	966	966	-	
Silver	5,437	5,494	57	@ 96% i.o. 95%
Treatment Charge	**(158)**	**(85)**	**73**	
Treatment Charge CIF China	(123)	(50)	73	long term vs spot
FOB Manzanillo to CIF	(35)	(35)	-	
Refining Charge	**(368)**	**(248)**	**120**	
Silver	(367)	(247)	120	@ $0.80/oz I.o. $1.20/oz
Penalties	**(42)**	**(42)**	**-**	
OME Fees	**-**	**(10)**	**(10)**	
Ag/Au concs (> 10k dmt)		(10)	(10)	0.16% of total revenue
Total offer per dmt	**6,002**	**6,242**	**240**	
Difference ($ per dmt)		**$240**		
Difference (%)		**4%**		
Quantity (dmt)		**20,000**		
Total annual net increase in revenue		**$4,796,579**		

OPENMINERAL

OME summary of the progress up to date

- OME product fully functional, > 100 registered clients, good level of activity
- Constantly improving and adding new features: Smart Calculator, About Company Section, RFP module
- Multilingual customer support and communication plan in place
- Growing team: adding expertise in shipping, finance, strategy, sales, strengthening local presence in China
- Closing collaboration agreement with new partners for trade finance, shipping, insurance, and assaying services

OPENMINERAL

Future vision for Open Mineral

1. To create an ecosystem of players which provide the most secure, reliable and efficient solution for concentrate buyers' and sellers' deal execution, powered by blockchain technology

2. To drive digitization and transformation of the global mining sector through Mining Alliance

3. Perfect basis for setting up Mining Alliance and introducing distributed ledger technology for contract execution, trade finance and logistics

OPENMINERAL

2. To create an ecosystem of players involved in deal execution powered by blockchain

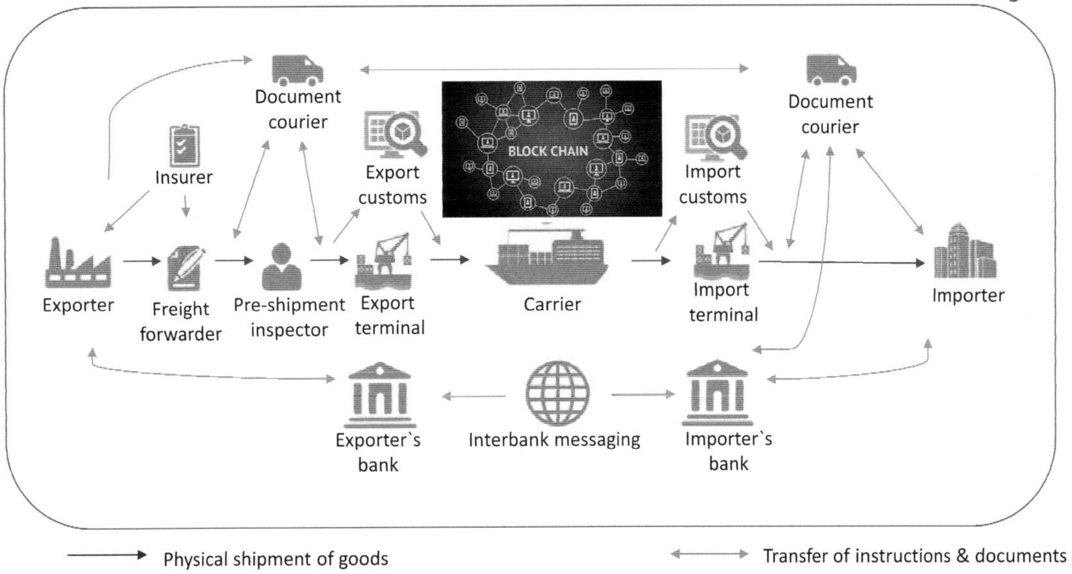

OPENMINERAL

14

3. Catalyst of the Mining Alliance

- Drive digital transformation of the Mining Industry
- Be a catalyst for the adoption of the blockchain technology by creating pragmatic solutions to existing painpoints: inefficiency, lack of trust, manual paperwork, etc.

OPENMINERAL

14

Next Steps: building a pilot logistics chain powered by Blockchain

3 clear cases for Blockchain for concentrate trading with proven value added

Bills of lading

- ✓ Reduced administration and operational costs
- ✓ Error-free documentation and fast transfer of original documents.
- ✓ All authorized parties can issue, transfer, endorse and manage shipping and trade related documents through our secure decentralized network.

Letters of Credit

- ✓ Process streamlined, made faster and more efficient
- ✓ Information instantly updated and visible to all users
- ✓ Avoid documentary fraud and mitigate risks
- ✓ Secure transfer of value
- ✓ Unlocking capital, which otherwise would be tied up

Assay Exchanges:

- ✓ Assay exchange results are directly submitted via smart contract.
- ✓ Results are instantaneous and available to all parties as soon as they are uploaded.
- ✓ No tampering or discrepancies - Results encrypted with unique hash.
- ✓ Both parties can rely on totally accurate and fair results from surveyors.

Set up for the pilot

- 1 mine and 1 smelter
- 1 shipping company or
- 1 bank ready to provide LCs for a list of select buyers
- 1 blockchain technology partner

15

OPENMINERAL
Thank you for your attention
For more information and to register, please visit our website
www. openmineral.com

Market Opportunities for Zinc Thermal Spraying

Frank Goodwin
International Zinc Association

What is Zinc Thermal Spraying?

Metallization of (steel) substrates

Also in the field: NJ DOT

Zinc Thermal Spraying processes

Twin-Wire Electric Arc Spraying

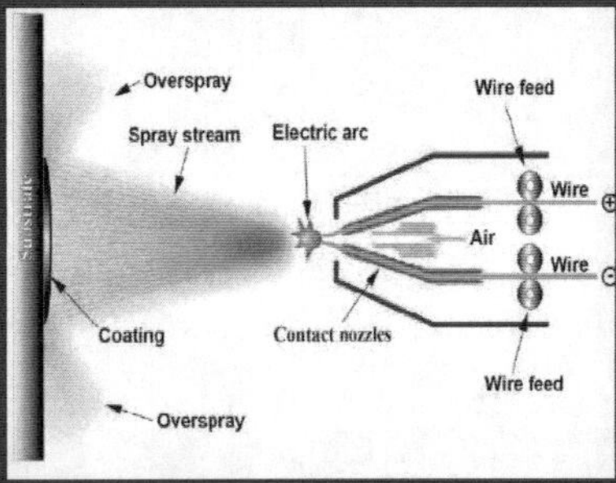

Flame Spraying

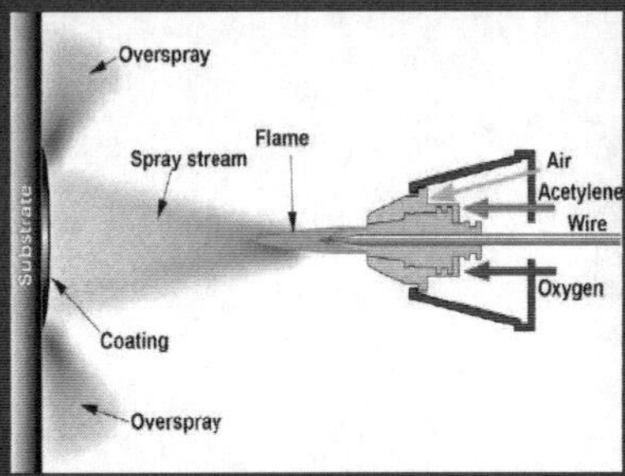

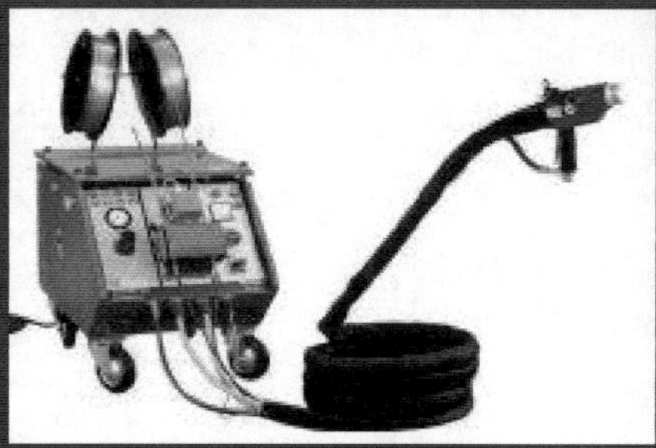

Thermal Spray Coating

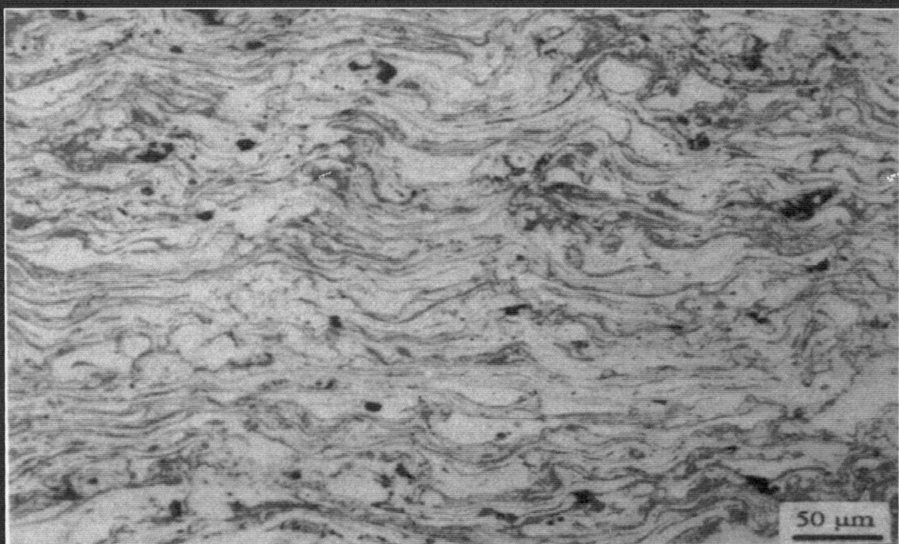

Extra protection
with
Duplex Systems

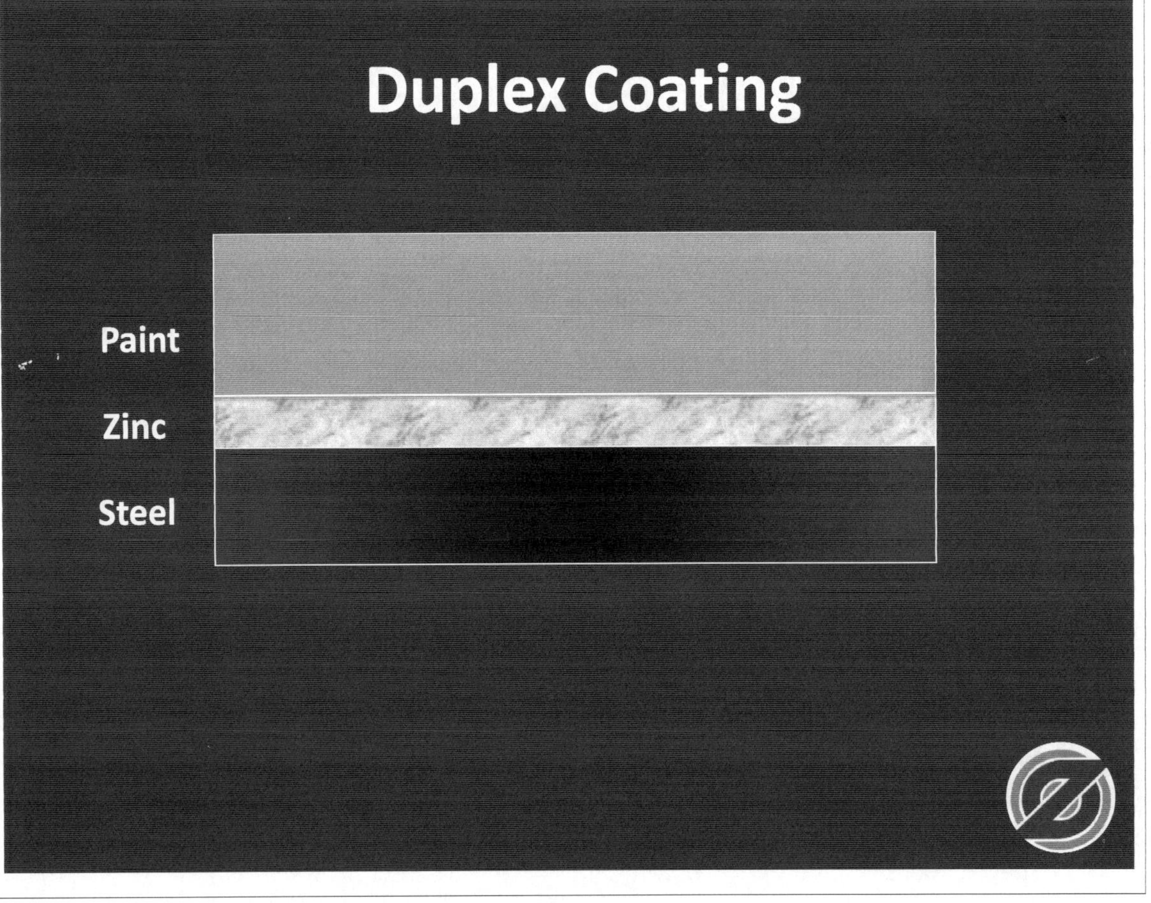
Duplex Coating
Paint
Zinc
Steel

Duplex vs Paint Coating

Duplex coating | **Paint coating**

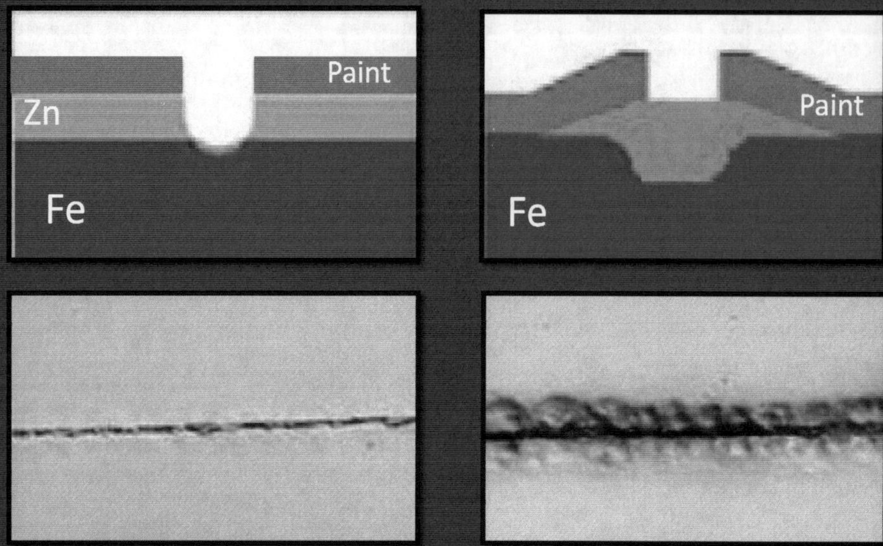

Duplex System Synergy

C5 environment

Duplex Service Life =

Factor x (Zn Life + Paint Life)

e.g. 1.5 x (15 years + 5 years)

= 30 years to 5% rust

Synergistic Factors

Industrial	1.8 - 2.0
Marine (immersion)	1.5 - 1.6
Non-aggressive climate	2.0 - 2.7

J.F.H van Eijnsbergen, Duplex Systems, Elsevier, 1994

Specifier Guidance (ISO 9223 & 12944)

Exposure type:	Life to first maintenance: Zn coating	Life to first maintenance: Duplex coating
C1 – dry	> 200 years	> 200 years
C2 - occasional condensation - Rural inland	>170 years	> 200 years
C3 - high humidity, some air pollution - Urban inland or mild coastal	42 – 170 years	>200 years
C4 - swimming pools, chemical plant - Industrial inland or coastal	21 – 42 years	42 – 84 years
C5 - Industrial with high humidity - High salinity coastal	10 – 21 years	18 – 37 years

Case Studies

- Buildings & constructions
- Bridges
- Railway
- Sports facilities
- Wastewater
- Ports
- Off-Shore
- Outer Space

"Rombak" Bridge

- Over the Fjord, Norway
- Corrosion Class C4
- Opened, 1964
- Duplex coating, 1970

2009 Inspection: "No corrosion was found after 39 years in service"

In Norway 2000+ steel bridges are zinc coated by standard

(Knudsen, Bjorgum, & Dossland, 2012; Photos: R. Klinge, NPRA, Patrik Engman)

"Älvsborg" Bridge

- Gothenborg harbor area, Sweden
- Corrosion Class C4
- Opened with duplex coating, 1966
- 2x patch maintenance of thin paint layer

2014: Original zinc spray coating will soon have its 50 years anniversary

(Experiences with ZTS Duplex coatings, Knudsen, 2014)

"Vilsund" Rail Bridge

- Rail bridge in Denmark
- Corrosion Class C5
- Lower parts in splash zone of salt water
- Opened with duplex coating, 1939
- Riveted steel construction with overlapping joints (not corrosive friendly)

1979 after 40 years: blast cleaned to steel. New TSZ layer and alkyd top coating.

2013 after 34 years: due to incompatible organic top coating, cleaned to steel again. New TSZ layer and modern top coating

(Experiences with ZTS Duplex coatings, Knudsen, 2014)

Bridge of Lions

- St Augustine, FL
- Opened, 1927
- Full rehabilitation, 2006-2010

- Zinc thermal spray and paint structural steel
- Galvanize and paint handrails and tower steel

MATERIALS PERFORMANCE, NACE Int, Vol. 49, No. 11, Nov 2010

Castleton, IN

I-69 over East 82nd St
1970 - NB lanes HDG
- SB lanes paint

- 2002 Widened
- Zinc thermal spray SB lanes

- 2017 Inspection: excellent condition

1995 Rehabilitation of Staten Island Ferry terminal, New York City Harbor

International MetalFusion Corp., www.metalize.net

The result of a 10 Year Inspection

"No corrosion on any of the metallized and coated surfaces"

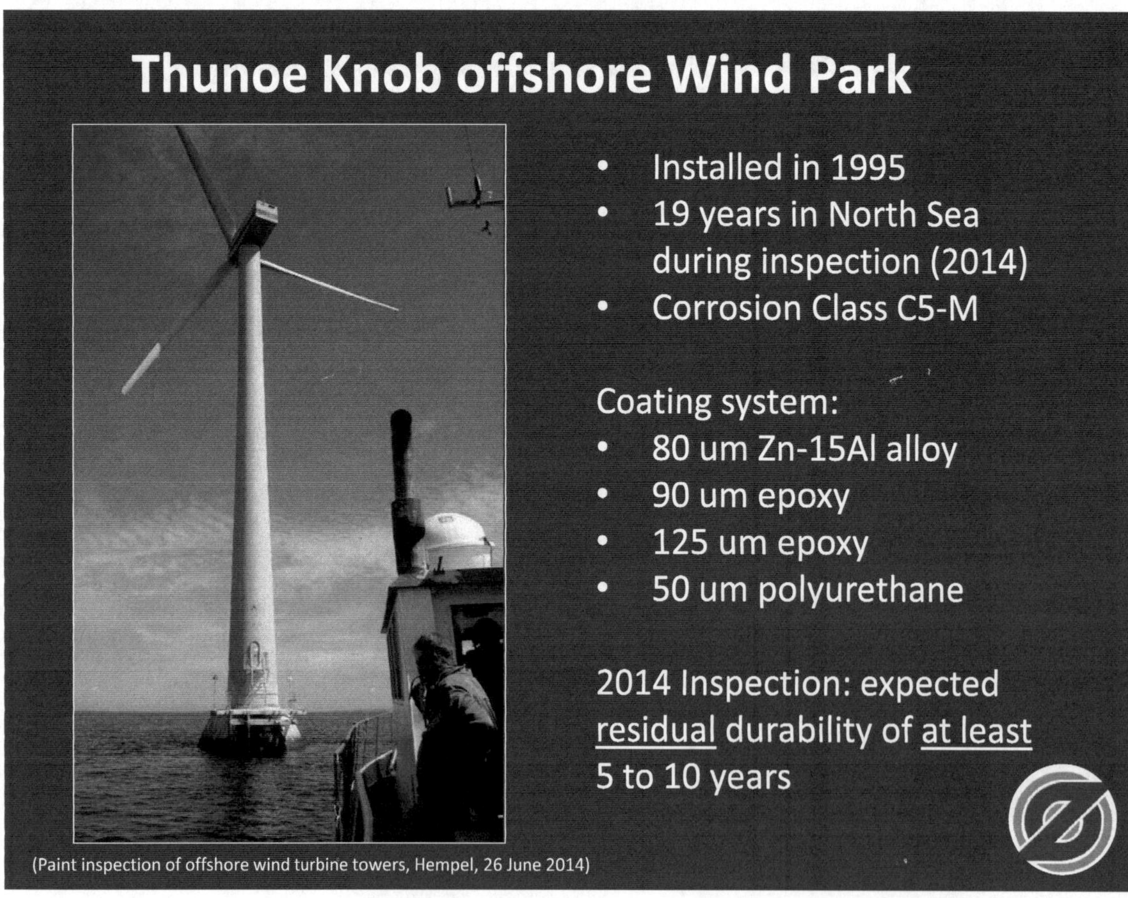
Thunoe Knob offshore Wind Park
• Installed in 1995
• 19 years in North Sea during inspection (2014)
• Corrosion Class C5-M
Coating system:
• 80 um Zn-15Al alloy
• 90 um epoxy
• 125 um epoxy
• 50 um polyurethane
2014 Inspection: expected residual durability of at least 5 to 10 years
(Paint inspection of offshore wind turbine towers, Hempel, 26 June 2014)

Tower 005: On all tower sections no signs of rusting or other defects.

Tower 005 door frame:
In 6 o'clock area, with generous layer of bird manure with a high time of wetness must have provided a high corrosion stress, yet not signs of breakdown.

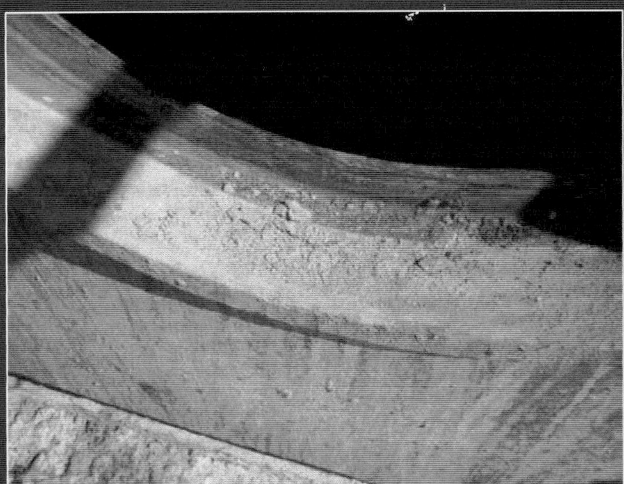

Tower 005: The antennas for monitoring system are preferred resting zones for the cormorants, so are covered in manure.

Zinc anode for concrete protection

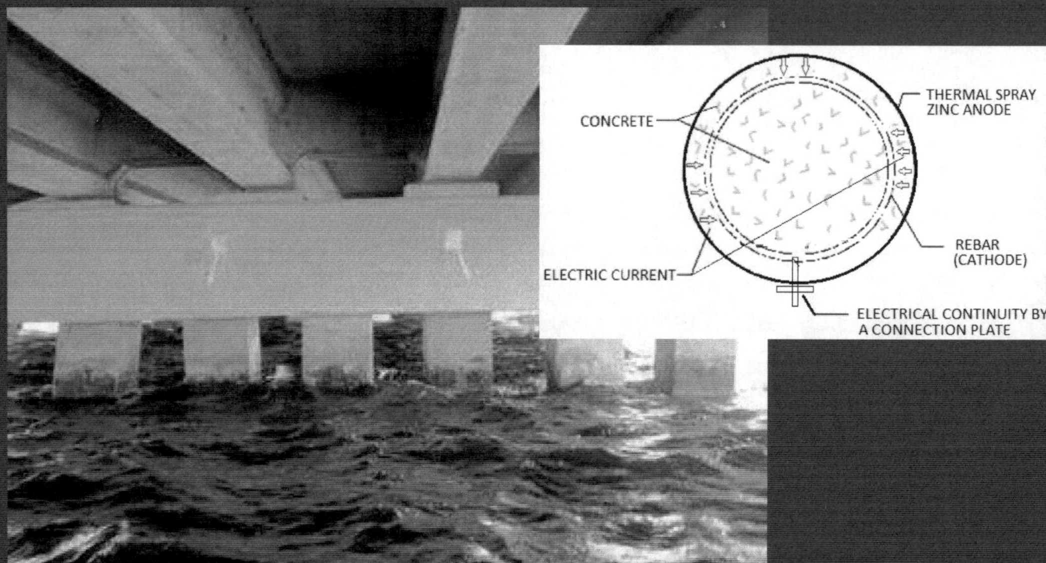

Application of sprayed zinc anode on concrete on a structure in a marine environment.

SPRAYTIME, 4th quarter 2016, volume 23

Safe use of corrosion protected gas bottles

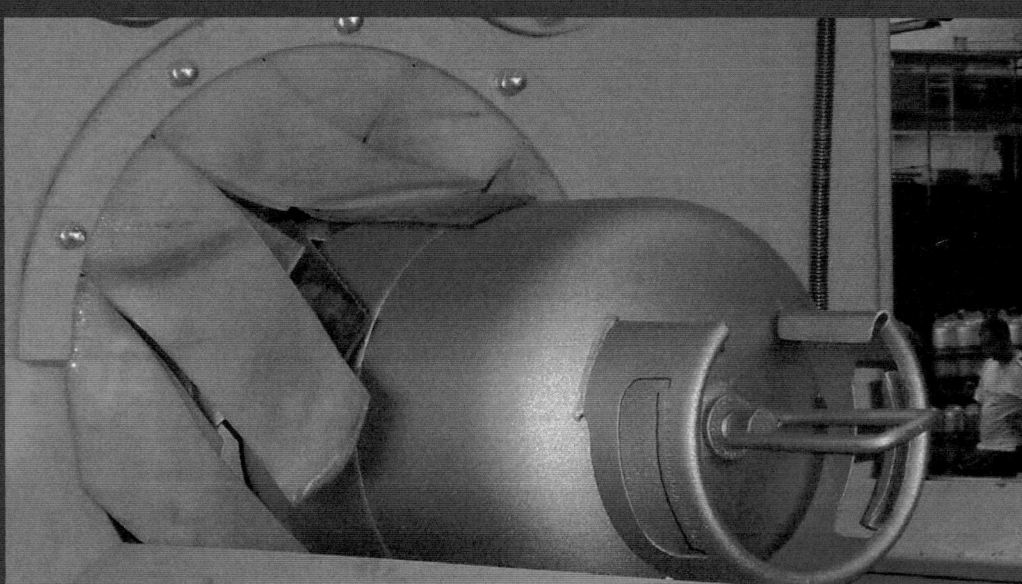

www.metallisation.com

Life Cycle Assessments

Life Cycle Assessment

Total cost of ownership of a wind turbine in comparison

Relative Life Cycle Cost (%)

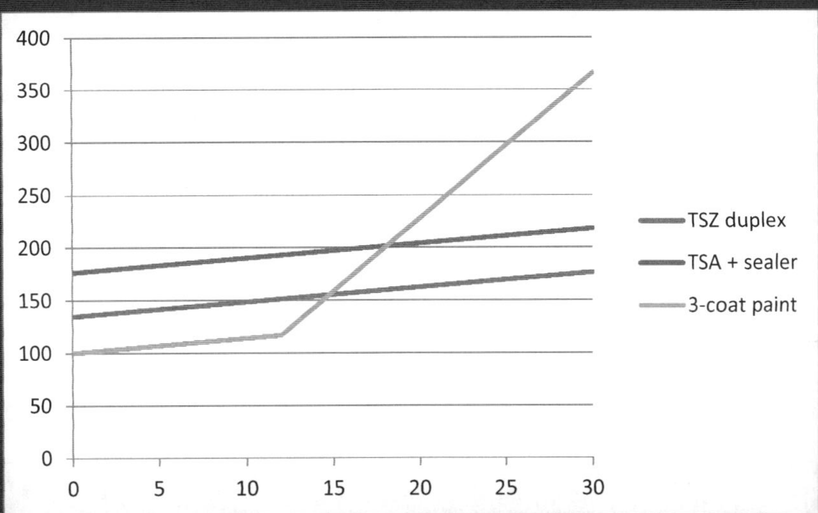

Construction Lifetime (Years)

"Low Maintenance Coating Systems for Constructions with Long Lifetime"; Ole Øystein Knudsen, SINTEF

Life Cycle Assessment

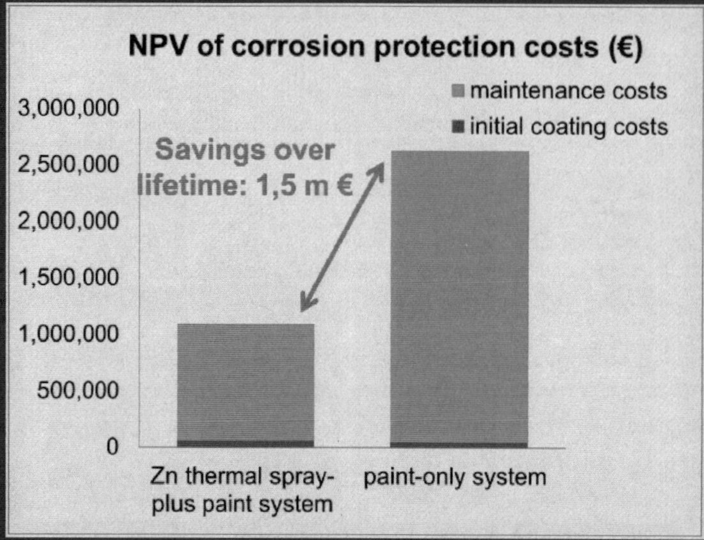

Savings are 0.8 €cents per KWh for each KWh generated over the 20 year life of the structure for a 3.6 MW wind turbine with a capacity factor of 0.4

"Cost Savings in Corrosion Protection for Offshore Wind Energy Installations"; Frank Goodwin, IZA, 2013

Life Cycle Assessment

Capital and Maintenance cost of 2 comparable bridges in Norway (in NOK/m^2)

Table 3. Life cycle cost comparison of Brevik and Rombak bridge.

Rombak bridge		Brevik bridge	
Original coating	590	Original coating	390
Topcoat maintenance 2012	260	Patch maintenance 1970s	310
		Patch maintenance 1980s	310
		Total rehabilitation 1992	500
		Patch maintenance 2013	310
SUM	**850**	**SUM**	**1820**

Duplex Coating (1964-2014)

Paint Only Coating (1962-2014)

(Experiences with ZTS Duplex coatings, Knudsen, 2014)

Market and Market Opportunities

2017 estimated Thermal Spray Markets

- Europe: 40,000 mt or 88.2 Mlbs /yr
- India: 15,000 mt or 33.1 Mlbs /yr
- China: 30,000 to 40,000 mt /yr or 66.2 to 88.2 Mlbs /yr
- USA: 5,000 mt or 11.0 Mlbs / yr
- **Total: 90,000 to 100,000 mt / yr or 198.5 to 220.5 Mlbs / yr**

Opportunity - Protection of rail tracks

Use of thermal spray zinc coated rail in aggressive rail conditions such as coastal track, wet tunnels, level crossings, salt pans and stray current environments.

British Steel Rail, ZINOCO technical datasheet & courtesy IZA

Protection of ships by Thermal Spray Zinc is nothing new

DECKS OF 90 FT BARGES,
Operated in Bristol Channel by T R Brown & Sons Ltd.

Case History No. 5

Environment
Severe marine, constantly exposed to salt water spray.

Date of Treatment
1962–1964

Protective Treatment
The welded decks of ten barges, heavily pitted and totally perforated in places, were repaired and sprayed with 250 μm (0.010 in) zinc, sealed with one coat of aluminium – pigmented unmodified phenolic resin sealer.

Maintenance
At 3-4 year intervals one coat of ordinary red oxide paint is applied for decorative purposes only.

Inspection
1974 (10-12 years)
Zinc coating intact on all barges. At one spot 3in – 4in dia. on the deck of one barge where the bargee's foot always lands and rotates as he steps out of the wheelhouse, the zinc coating has worn away to the steel but there is no active corrosive attack. Other steelwork not zinc sprayed, e.g. wheelhouse sides and hatch coverings, have rusted through.

Conclusion
Good resistance to certain types of abrasion and impact coupled with corrosion resistance enable sprayed zinc of reasonable thickness to protect ships' decks for long periods. The decks of many Icelandic trawlers have been successfully protected by metal spraying.

Left: A zinc sprayed deck after twelve years service. The largest area, of medium shade is the aluminium-filled phenolic original sealer. The very much lighter parts, spidery in appearance, are zinc metal where the sealer has been worn off by foot traffic. The darkest parts in the upper right hand corner of the photograph are traces of the red oxide one-coat repaint applied the year before this photograph was taken.

Barge Thermal Sprayed with Zinc in 1962

Courtesy EGGA

Still opportunity - Protection of ships

Use of thermal spray zinc reduces future coating application reworks (light patch-up), typically required up to 3-4 years after application rather than 1-2 years with paint only systems

Stuart Milton - www.metallisation.com

LCA of painted ships opens doors

Vessel (ship) life cycle cost analysis

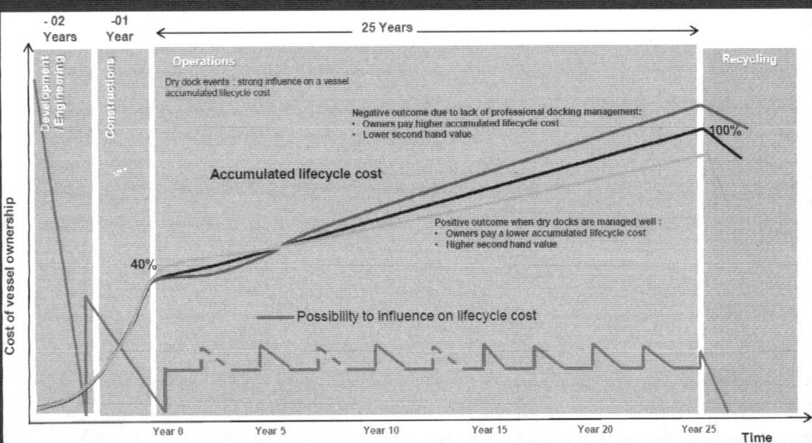

Accumulated life cycle cost when ship is in service is **60%** of total cost, mainly related to corrosion maintenance (currently mostly with paint systems only)

(O Dinu, Maritime vessel obsolescence, life cycle cost and design service life, Materials Science and Engineering 95 (2015)

Corrosion in the navy turn ships into not ready to battle

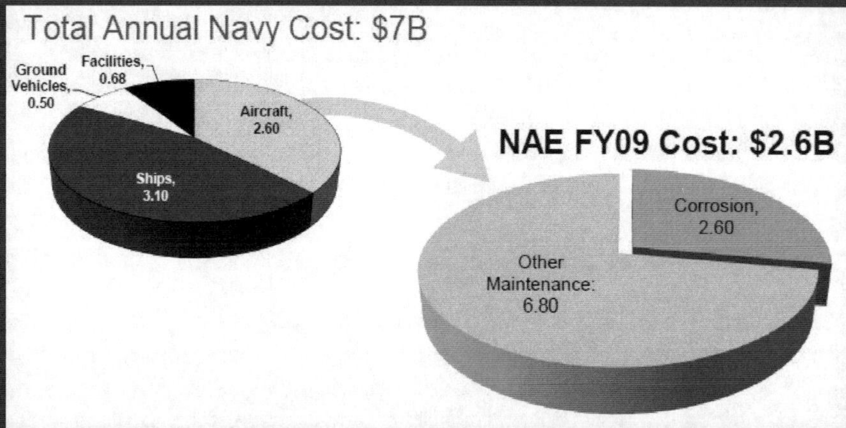

25% of DON maintenance and 38.2 days of a Ship availability are expended on Corrosion

Mr. E. Dail Thomas II, Consultant to the U.S. Navy Corrosion Executive

Offshore Wind expanding to USA and China

Coatings Industry News

Main News Page

Firm Bids to Build MA's First Offshore Wind Farm

More

Thursday, January 4, 2018

More items for **Program/Project Management**

More items from **North America**

In response to Massachusetts' recent request for proposals, Bay State Wind has put in a bid to build the state's first offshore wind farm.

Bay State Wind, which is a partnership between Ørsted (formerly DONG Energy) and Connecticut-based energy company Eversource, is slated to build a commercial-scale wind farm, and has also pledged additional incentives.

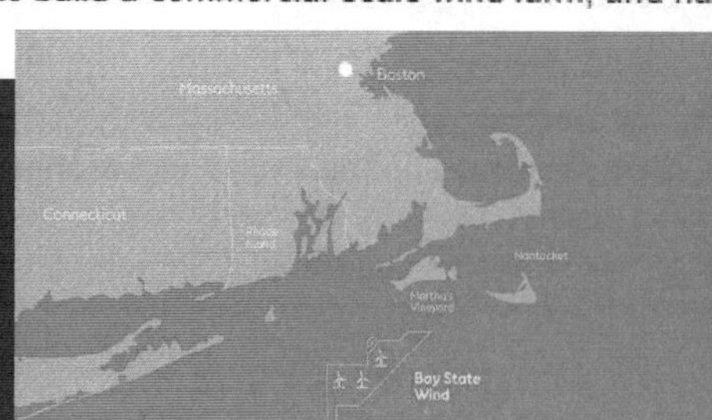

www.paintsquare.com

Can we get the French interested?

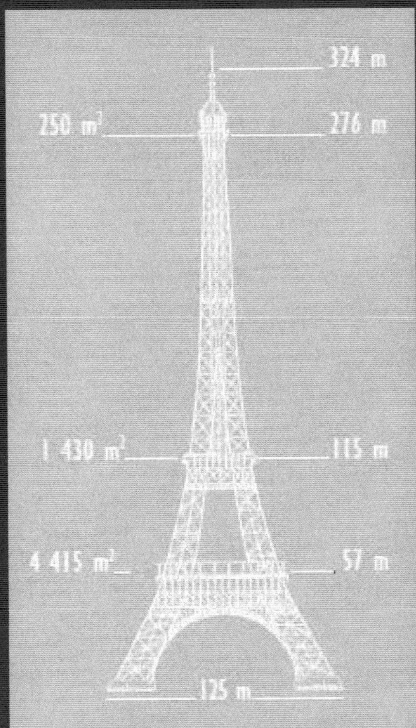

The Eiffel Tower has **NO** duplex coating (yet)

Key figures behind the painting of the Eiffel Tower:

- Weight of paint: approx. 60 mt
- Time required: 15 to 18 months
- Frequency: entire re-paint every 7 years.
- 25 painters
- 18 times painted till 2001, in total **619 paint man-years !!!**
- Cost of the 18th painting in 2001: **3 million euros !!!**

All you need to know about the Eiffel Tower

Potential Market Impact

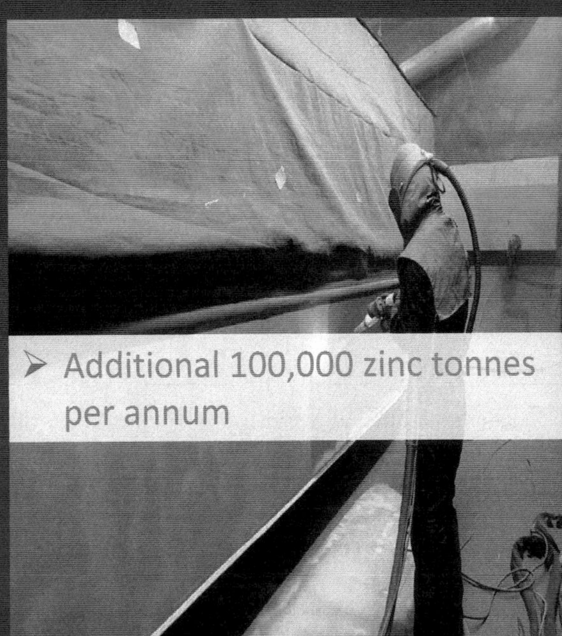

➢ Additional 100,000 zinc tonnes per annum

Summary

- Zinc Thermal Spray Coatings offer:
  - superior corrosion protection
  - longer lifetimes between maintenance
  - lower life cycle costs
- Many successful case studies
- Lots of potential in Bridges, Railways, Maritime, Navy and Renewable Energy

http://thermalsprayzinc.wp.zinc.org/

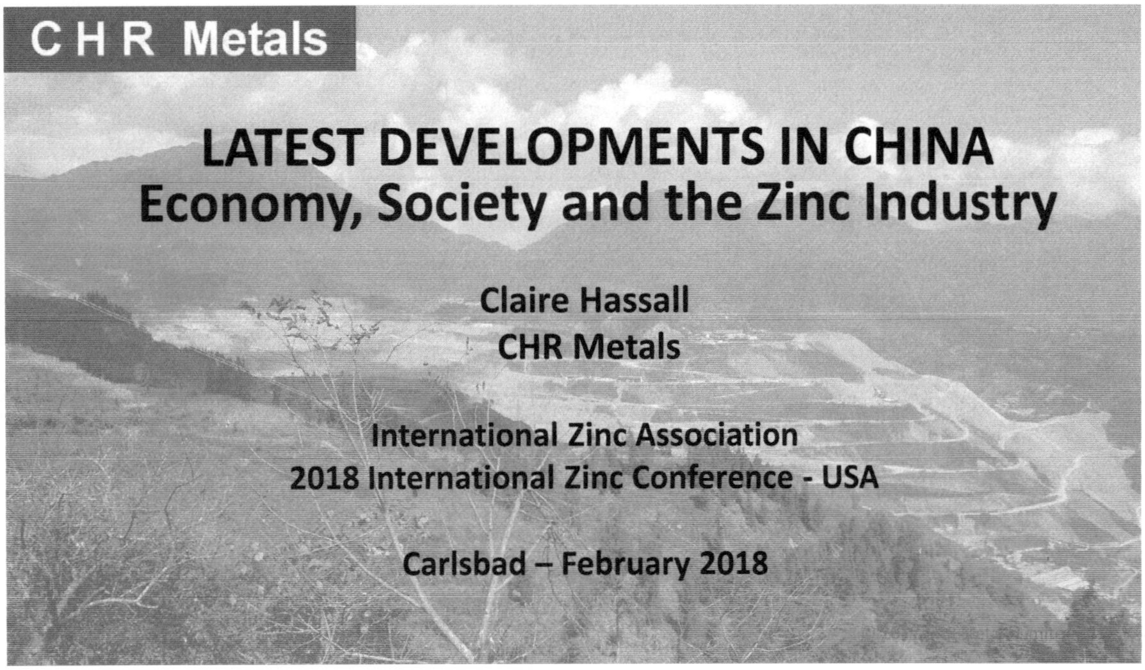

C H R Metals

Presentation Outline

- CHR Metals
- Economic developments in China in 2017
- Social trends in China
- Environment and safety
- Implications for Chinese zinc demand
- Chinese zinc supply-side developments
- Summary and outlook

February 2018 Slide 2

C H R Metals

Introduction to CHR Metals

- CHR Metals established in 2000 to provide independent, detailed analysis and forecasts of global lead and zinc industries
- Covering all aspects of mine and smelter supply and end-use consumption
- Data from original sources wherever possible
- A particular focus on Chinese market
- Offices in the UK and Xi'an
- Clients include producers, consumers, traders and hedge funds

February 2018 Slide 3

C H R Metals

2017 – a year of thirds

- Very strong growth in industrial production (IP) in Jan-Apr, with some slowdown May-Aug (8.9% y/y to 8.4% y/y)
- Slump in IP growth in final 4 months (+1.3%y/y) – but with output in Sept-Dec falling by 5.2% compared with May-Aug
- Legacy of 2016 stimulus, and strong private sector investment but these influences weakened as 2017 progressed
- Central government announced stringent measures to control pollution:
  - 19th CPC National Congress (October 2017)
  - Industrial shutdowns to prevent a repeat of 2016/2017 winter pollution in north China
- Industrial output surged before September but has subsequently slumped

February 2018 Slide 4

C H R Metals

19th CPC National Congress outcomes

- Emphasis on CPC governance
- Less emphasis on growth targets
- One belt-one road included in the CPC constitution
- Greater emphasis on the well-being of the people
  - Rural poverty alleviation
  - Improved universal education
  - Healthcare and health insurance
  - Environmental protection and sustainable development
  - Quality of employment and growth in earnings
  - Multi-tiered social security for urban and rural residents
  - Meeting housing needs

February 2018 Slide 5

C H R Metals

C H R Metals — Key drivers of growth are changing

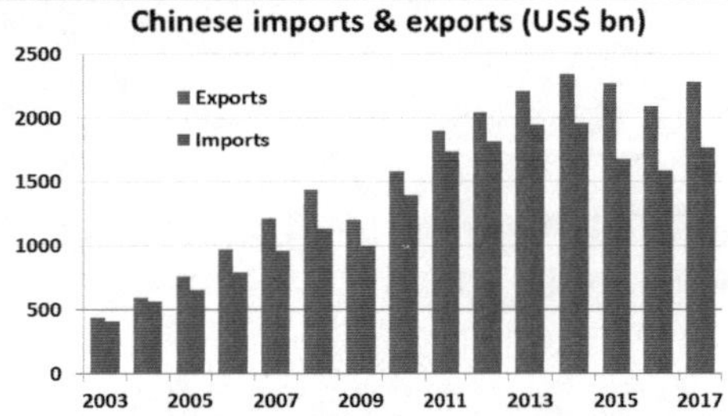

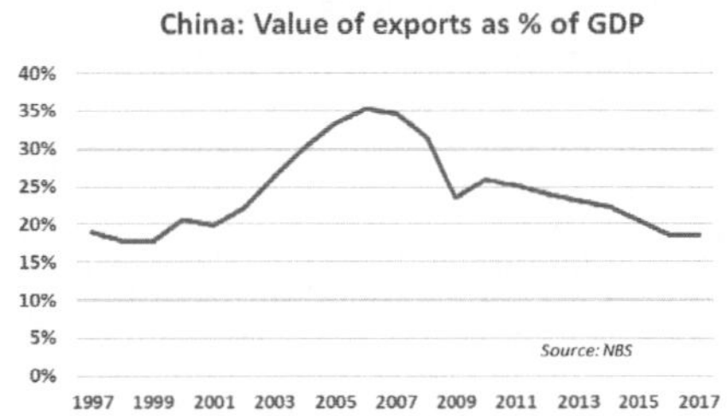

Source: NBS

- Exports remain important but no longer a key engine of growth for China, in spite of increase in $ value in 2017 (+7.9%)
- Value of China's exports (in US$) peaked in 2014, as did imports
- Value of exports as per cent of GDP peaked at 35% in 2006, but is now back to level seen pre-WTO accession (2017 =~18.5%)
- Domestic costs are rising – wages, land, environmental costs, etc.
- China is no longer as competitive as the "world's factory"

February 2018 Slide 7

C H R Metals — Key drivers of growth are changing

China: Breakdown of contribution to GDP by

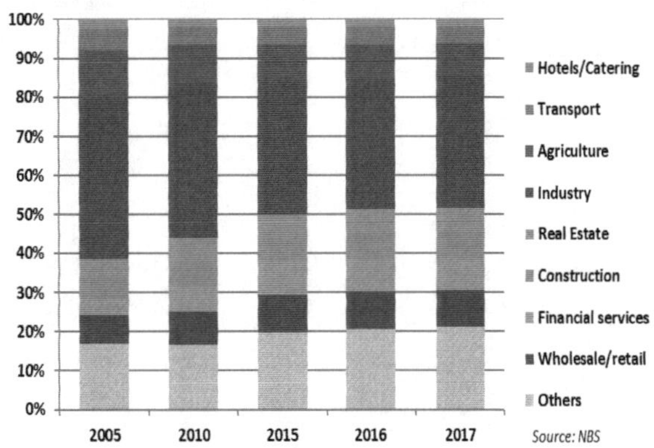

Source: NBS

- Share of industry in Chinese GDP has decreased from 41.4% in 2005 to 33.9% in 2017
- Industry's 2017 share increased slightly from 2016 level – prompted by the strongest growth in IP since 2012
- "Other" has been the fastest growing share in GDP ...
- ..and this trend is expected to continue
- "Other" includes health, education, recreation, IT, environmental protection, etc

February 2018 Slide 8

C H R Metals — Investment patterns show some changes

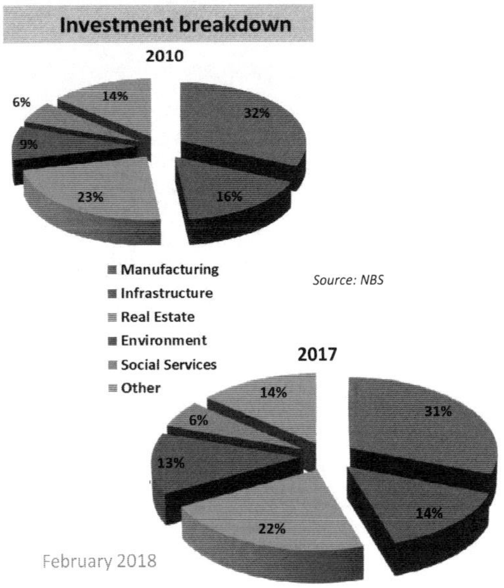

- Investment in manufacturing capacity accounted for 31% of total in 2017 – little changed from 2010
- Infrastructure investment (power, water & transport) saw small decline in share from 16% in 2010 to 14% in 2017...
- ...with the share of spending on environmental projects showing the greatest increase (from 9% to 13%)
- Surprisingly, no change in share of social services which includes investment in health and education
- Other (agriculture, mining, wholesale, hotels etc) unchanged

February 2018 Slide 9

C H R Metals — Concern about China's debt

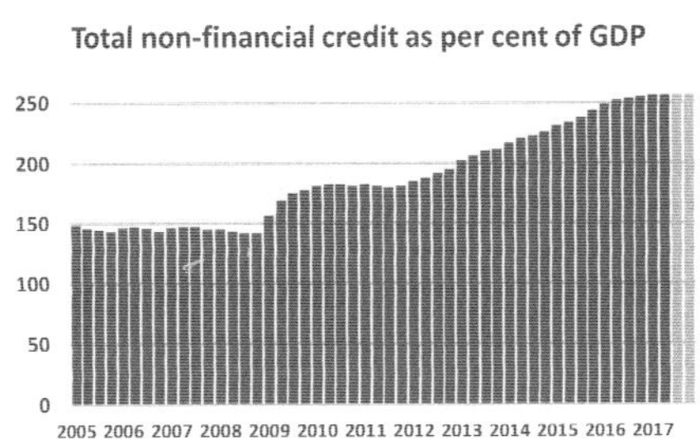

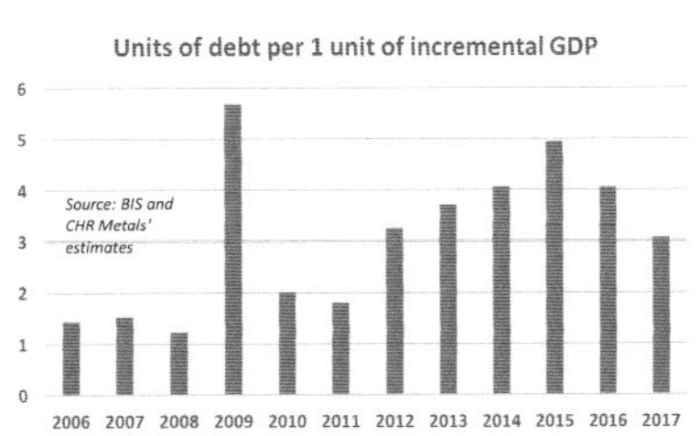

- Central government now appears not to be as focused on reducing absolute debt levels relative to GDP, seeking rather to reduce the risks associated with high debt levels
- Closer scrutiny of value and affordability of projects ...
- ..leading to cancellation/delay of some - e.g. reducing provincial plans for new general aviation airports, cancellation of subway in Baotou, Inner Mongolia & halt to work on 3 road projects in China-Pakistan Economic Corridor
- Superficially efficiency of debt in generating GDP growth has recovered but assisted by large-scale conversion of debt to equity by banks

February 2018 Slide 10

C H R Metals — Consumer spending is expanding – but boosted by credit

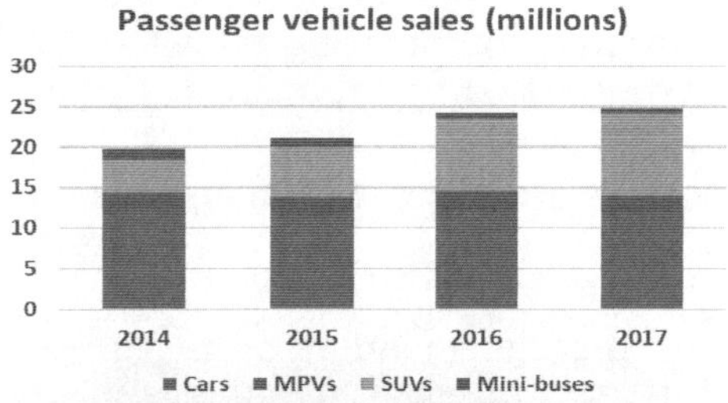

- China has a very large and expanding middle class keen to spend, and the Chinese consumer accounted for almost 59% of growth in 2017
- Retail sales rose by 10.2%, but online shopping rose 32.2% accounting for almost 20% of the total, up from 15% in 2016
- Increase in auto sales modest however as incentives were reduced
- Very large market for restaurants, entertainment; overseas travel is booming (+13.9% increase in expenditure on overseas travel in 2017)

February 2018 Slide 11 © 2018 CHR Metals Limited

C H R Metals — Social and demographic trends

- After Deng Xiaoping's reforms, rural migration provided a vast pool of cheap labour which led to the rapid expansion of Chinese manufacturing capacity and output, exports, infrastructure development, real estate development etc
- Lewis turning point – where there is no longer surplus rural/agricultural labour – balance of expert opinion is that China has passed this turning point, perhaps as long ago as 2010
- Impact seen in rising labour costs and rural depopulation
- Hukou system – preference for urban vs rural registration changing
- Previous one-child policy is leading now to a fall in the working age population in China, exacerbating labour cost pressures
- Changes in one-child policy have not led to the expected increase in the birth rate

February 2018 Slide 12 © 2018 CHR Metals Limited

C H R Metals

Changes in intensity of zinc usage

- Net additions to infrastructure will be on a downward trend – negative – but rising labour costs should promote more consideration of life-time project costs and use of galvanizing - positive
- Loss of low-end export markets to lower cost economies – negative
- Belt and road increasing Chinese funded projects – positive (particularly where projects would not otherwise happen) but ultimately may be negative for Chinese zinc demand as recipient countries develop manufacturing capability
- More service based economy (social care, health, education, travel etc) - negative
- Greater mechanisation and investment in Chinese agriculture – positive
- Growing consumer demand for durables/autos – positive
- Real estate is the paradox: surplus inventory but still more being built: high-end has been more zinc intensive than more standard construction which uses little zinc ...
- ..but improvements in electro-plating technology for aluminium die-castings since 2010, and increase in zinc price over last 2 years has seen major loss of market for zinc in die-castings – NEGATIVE

February 2018 Slide 13

C H R Metals

Focus on pollution impacting zinc demand now

- Serious pollution in northern China in winter 2016/17, along with strengthened commitment to environmental protection, has led to much more stringent controls on industry in northern provinces since late September
- Plant closures and reduced operating rates; temporary vehicle usage bans based on license plate numbers in some cities; conversion of industrial boilers from coal to gas etc
- Cold winter in northern China in 2017/18 led to higher demand for power and for gas; shortage of gas supplies led to further industry closures as consumers given preference over industry
- Galvanizing lines and other zinc consumers in affected areas subject to enforcement of environmental standards and to temporary closures
- Air quality in northern China significantly better this winter than a year ago

February 2018 Slide 15

C H R Metals

Environmental checks limiting zinc output?

- 2016 and 2017 have seen headline news of mines being shut because of official checks on compliance with regulations
- Some mining areas closed completely because they are in areas of outstanding natural beauty and importance; zinc production affected not large e.g. Gongcheng in Guangxi, Qingling Mountains in Shaanxi
- Compliance checks continued in 2017, but in reality extensive temporary closures were limited to just one area in western Hunan with some small losses also in Sichuan
- State-owned mines operate while checks taking place; other mines that had already met all environmental and safety requirements in previous checks also able to continue operating while checks took place last year
- New mine starts and expansions outweighed loss of output from mine closures in 2017; re-organisation and consolidation of mining licenses supporting output growth in 2018
- Zinc smelters largely unaffected by environmental regulations; exceptions include Zhuzhou which will relocate some time during the next 2-3 years, Heli Zinc in Sichuan closed at end 2016 and some small secondary producers in Hunan

February 2018 Slide 16

C H R Metals

Environment & safety in mining & dressing

- Mining areas have been subject to "control" to clarify rights and licenses with specific measures ordered to improve safety and environmental performance. Control has been exercised by the need to provide operating licenses to purchase explosives
- Environmental measures include modern tailings dams, with groundwater monitoring, water treatment plants, online monitoring of any discharge water, rehabilitation of waste rock dumps etc
- Six safety steps in mining:
  - Logging of underground workers
  - Underground refuges
  - Self-rescue sets
  - Surface to underground communications
  - Full visual monitoring
  - Escape routes

February 2018 Slide 17

C H R Metals
开发与
护同步

C H R Metals
铸牢质量基础 打造质量强企

C H R Metals

C H R Metals
细心、周密、认真、负责、爱岗

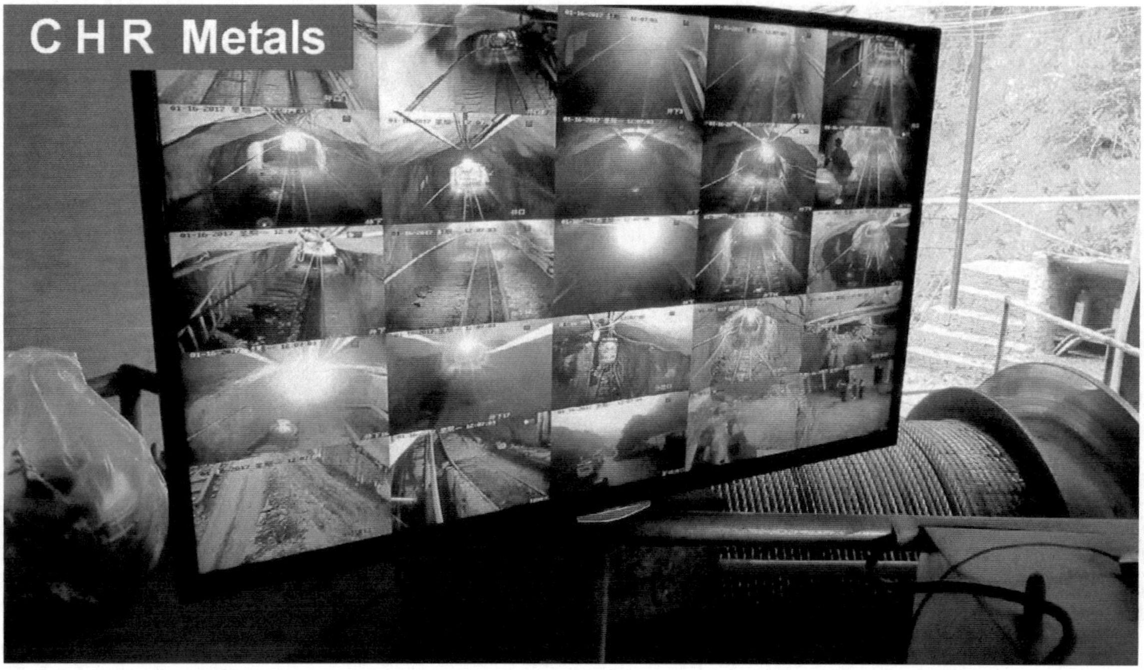
C H R Metals

C H R Metals

CHR Metals

CHR Metals

Smelting and treatment of residues

- Emissions standards have tightened and are enforced
- For example, central government has set SO_2 limit at 100mg/m^3 – considerably lower than in some other parts of the world
- Closure/relocation of operations has been, and is being, required in residential areas (Zhuzhou, Heli Zinc)
- More comprehensive recovery of by-products from residues – driven both by environmental regulations but also by straight economics, with higher main and by-product prices offering commercial opportunities to smelters
- Water treatment required and, in some locations, no water discharge permitted
- Environmental regulations hitting treatment of steel plant dusts

February 2018 Slide 26

C H R Metals

Closing observations

- Emphasis of CPC policy more on improving quality of life than on growth
- Economic policies directing huge spending on fixed asset investment changing with greater consideration of returns on capital expenditure
- More focus on containing risks implicit in high and expanding debt, whoever holds that debt
- Expect to see spending on construction and physical infrastructure as a share of GDP to fall, and for zinc demand to flat-line or fall in some years
- Expect to see rising zinc mine production as reforms lead to consolidation and the development of larger scale, more efficient operations helped also by improved infrastructure in more remote regions in far north and west
- Expect to see further modernisation in zinc smelting – larger units – but in the longer-term the elimination of smaller plants – with only modest increases in total capacity

February 2018 Slide 27

Zinc concentrate market outlook

Wood Mackenzie
A Verisk Business

Jonathan Leng, February 2018

Trusted commercial intelligence

woodmac.com

Global growth continues to be dominated by China

Consumption growth moderating in the medium-term

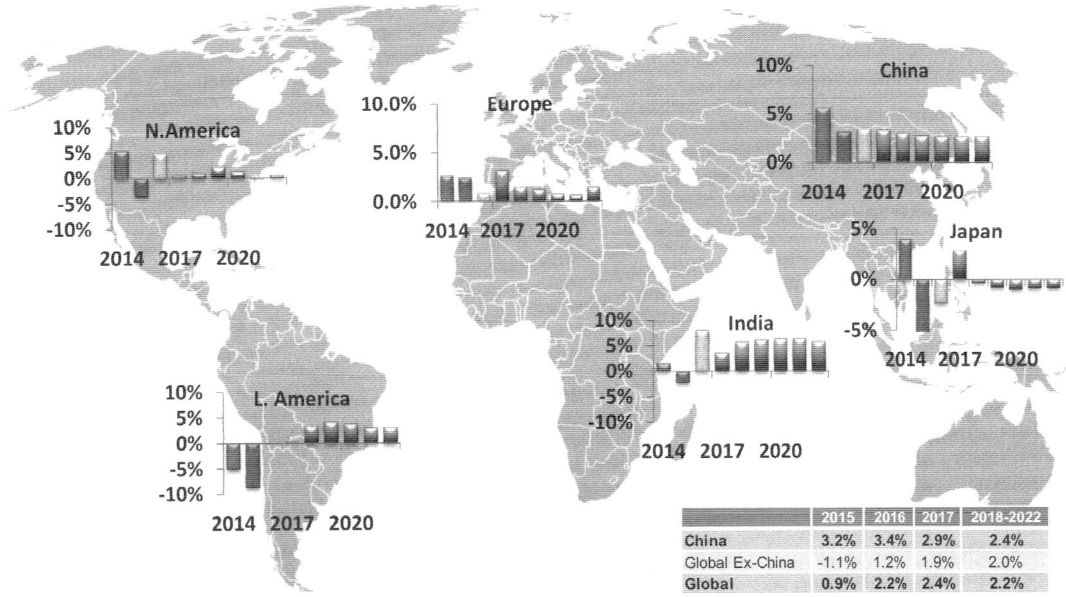

	2015	2016	2017	2018-2022
China	3.2%	3.4%	2.9%	2.4%
Global Ex-China	-1.1%	1.2%	1.9%	2.0%
Global	0.9%	2.2%	2.4%	2.2%

Source: Wood Mackenzie

The world needs new zinc mines

2016 saw the biggest fall in global mine output ever, with production almost recovering in 2017

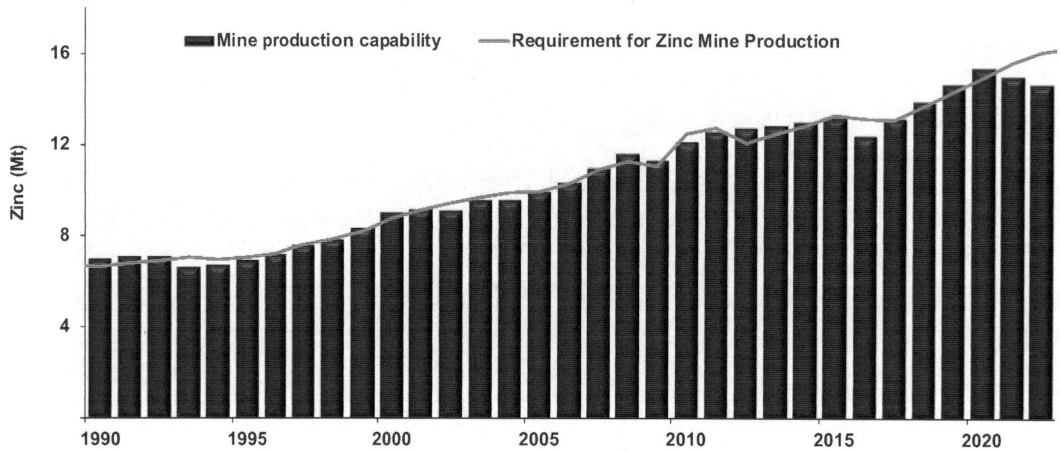

Source: Wood Mackenzie

Closures and production cuts of 1Mt Zn in both 2016 and 2017

Price-induced Zinc Mine Production Capability Losses (loss vs 2014 production kt Zn)

Mine			2015	2016	2017	
Century (500kt/a Zn)	Australia	Reserve depletion	73	450	465	Tails project from 2018
Lisheen (165kt/a Zn)	Ireland	Reserve depletion	30	130	130	Closed
Mae Sod (40kt/a Zn)	Thailand	Reserve depletion	7	7	42	Closed
Total planned			**110**	**587**	**637**	
Al Masane	Saudi Arabia	Early 2016		20		Resumed
Endeavor	Australia	H2 2016		35	20	Resumed
Middle Tennessee, Doe Run	USA	Late 2015		60	25	Resumed
Campo Morado	Mexico	Late 2015	21	22	21	To resume in 2018
Myra Falls	Canada	Mid-2015	18	27	27	To resume in 2018
Wolverine	Canada	Early 2015	38	38	38	Probable project
Total price/technical			**77**	**202**	**131**	
Iscaycruz (80kt/a)	Peru	Glencore	7	84	84	Probable project
Kazzinc (220kt/a)	Kazakhstan	Glencore	2	10		Resumed
McArthur River expansion (185kt/a)	Australia	Glencore		25	15	Probable project
Mount Isa (445kt/a)	Australia	Glencore		150	210	To resume in 2018
Total Glencore strategic			**9**	**269**	**309**	
Total			**196**	**1058**	**1077**	

Source: Wood Mackenzie

Disruption of concentrate trade to China

Supply cuts and mine closures disrupted trade flows. Miners and traders reduced exports to China

Displacement of Chinese concentrate imports 2015 to 2017

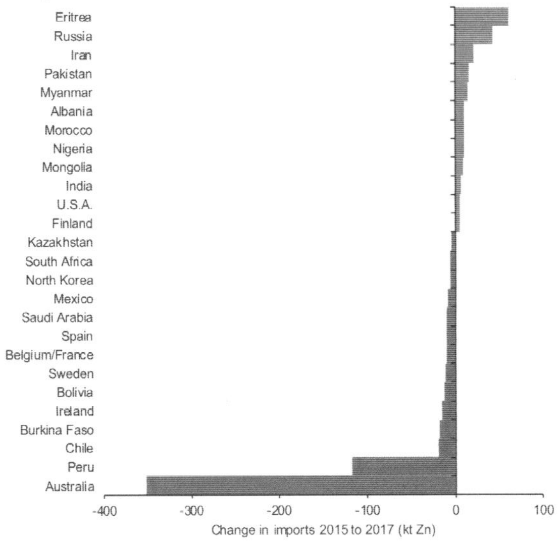

Source: GTIS, Wood Mackenzie

- ❑ Some concentrate shipments from Australia and Peru were switched from China to smelters in Europe, Japan and South Korea.
- ❑ This reflected the long-standing relationship between some mining companies, smelting companies and traders, buying under frame contracts and supplementing with spot
- ❑ Chinese smelters more reliant on domestic mines for feed. Imported concentrate tends to be purchased on short term contracts or from the spot market
- ❑ Several mines and traders increased shipments to China to take advantage of the low spot TC for imported concentrate. New mines Bisha and Duddar, Kyzyl and Noion Toloigsky

Mine closures and cutbacks have depleted concentrate stocks forcing spot treatment charges to record lows

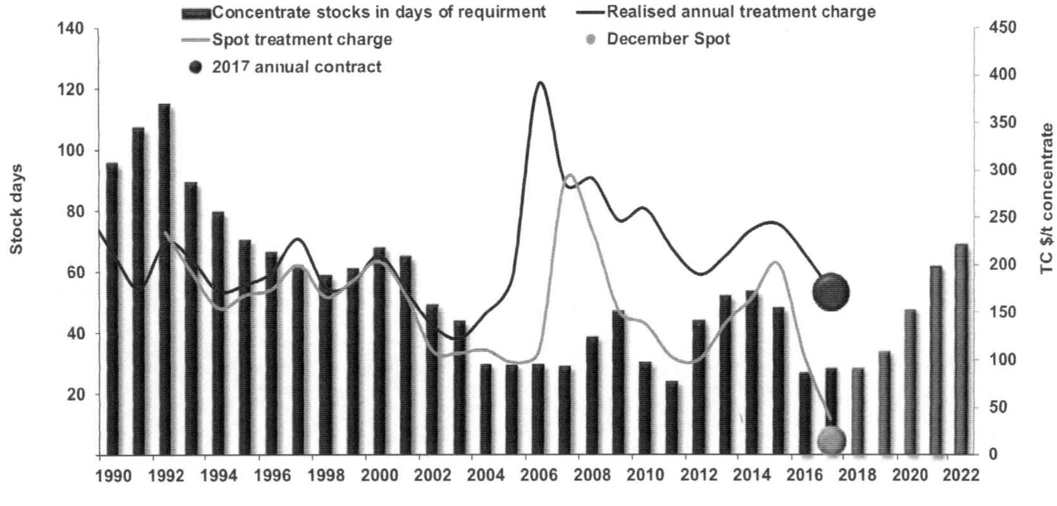

Source: Wood Mackenzie

Chinese smelters hit by low TCs but revenue has been supported by PP and free zinc

Price participation for domestic concentrate and free metal (plus the SHFE-LME price arbitrage at times) for spot TCs have supported smelter revenue

Chinese smelter revenue: Domestic conc. 100% Payability,80:20 price participation

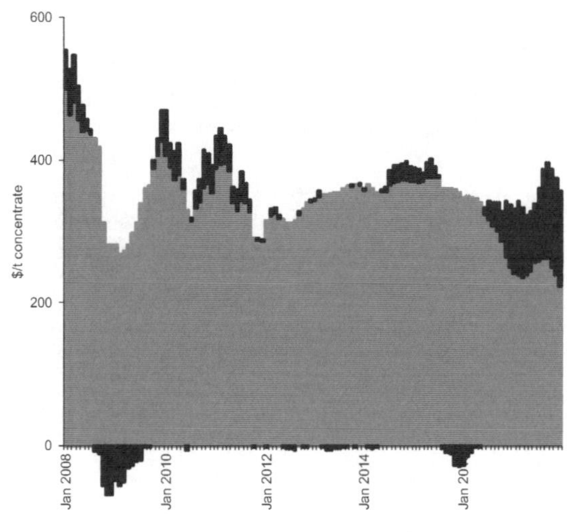

Source: Wood Mackenzie

Source: Wood Mackenzie

Chinese smelter revenue: Spot concentrate

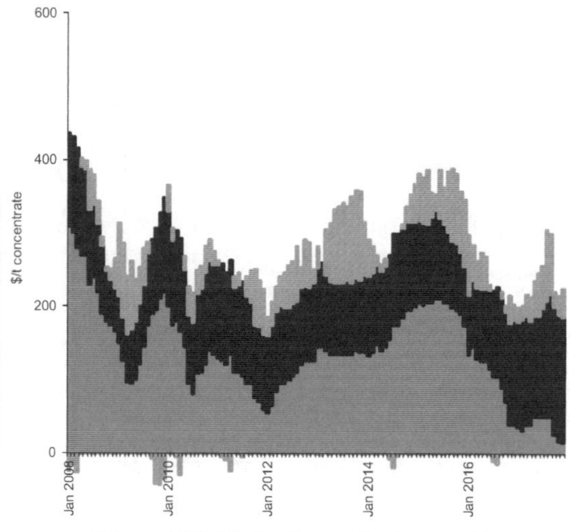

Source: Wood Mackenzie

Chinese refined production growth stalled in 2017, before rising sharply in Q4

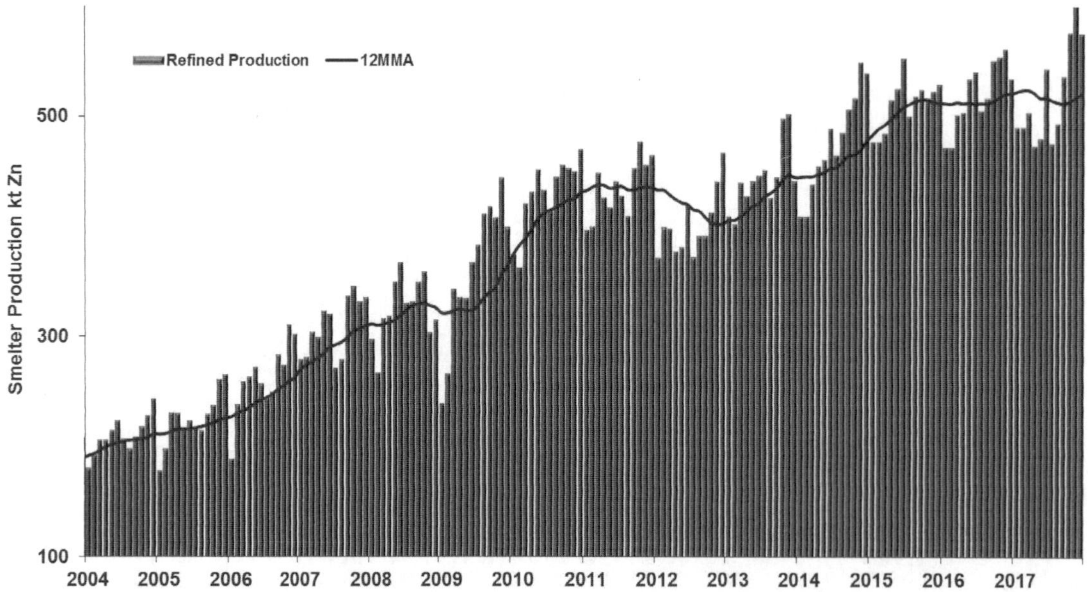

Source: CNIA, NBS, Wood Mackenzie

ROW smelter income at decade low (Realised TC + bonus metal + premia per tonne conc.)

Mine and smelter income per tonne concentrate (53.5% Zn, pay 85%, recover 96%)

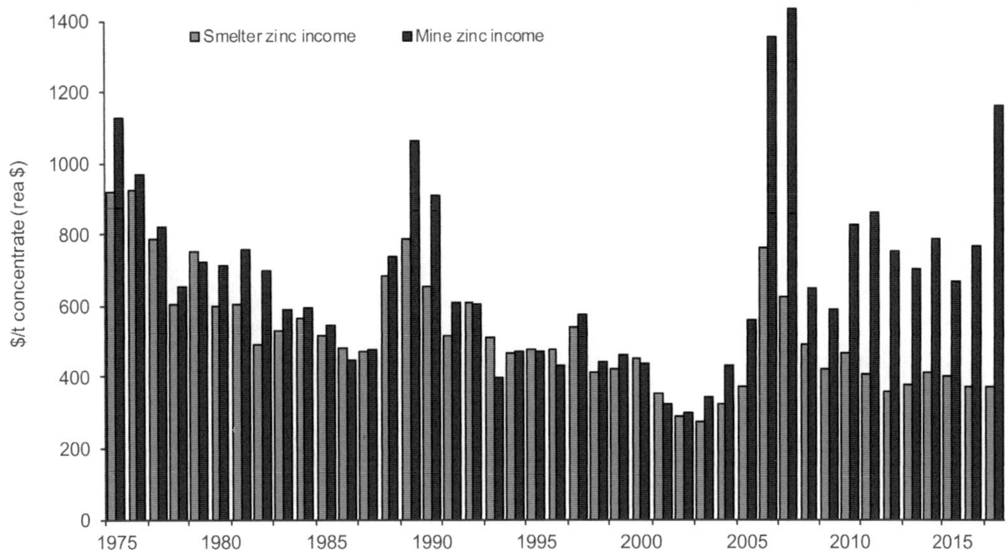

Source: Wood Mackenzie

The drop in treatment charges has increased the divergence between mine and smelter cash margins

Cash margin, recovered zinc

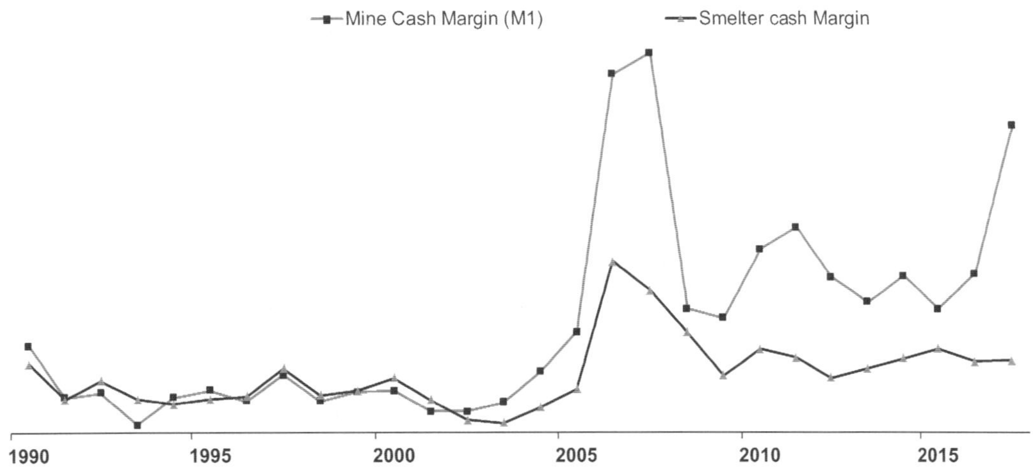

Source: Wood Mackenzie

woodmac.com 11

The need for new smelter capacity is critical

A shortage of smelter projects, plus cancellations and delays to existing projects is causing a smelter bottleneck in the near future

Global smelter capacity and requirement for new production

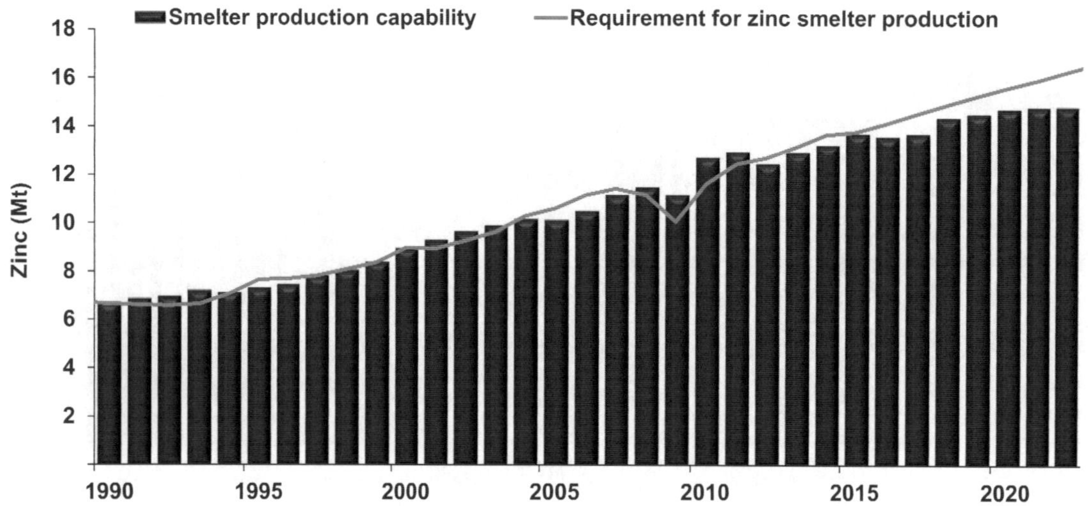

Source: Wood Mackenzie

woodmac.com 12

The world needs new zinc mines (2Mt/a Zn by 2022)

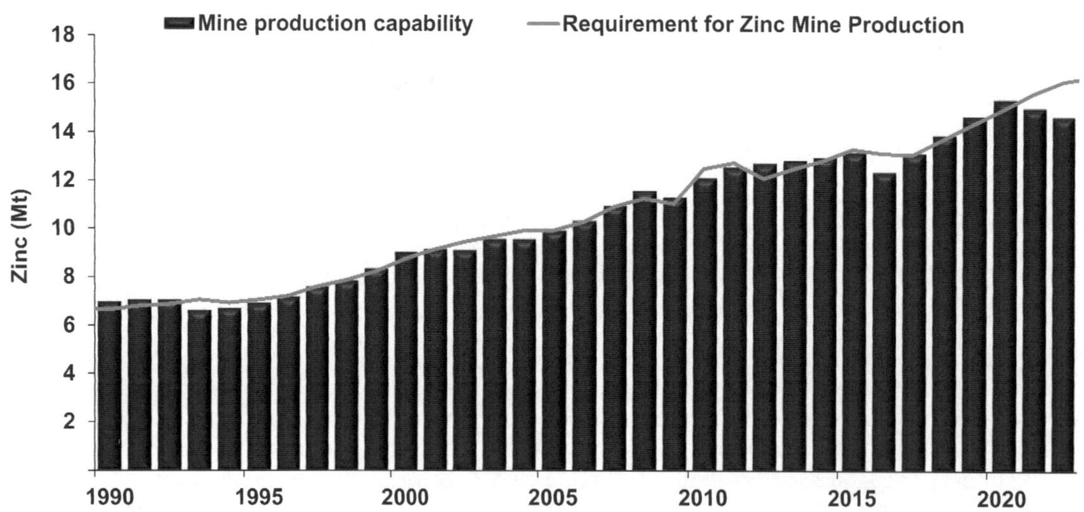

Source: Wood Mackenzie

Mining companies are gearing up for better times

Recent corporate activity

Glencore acquires control of Volcan, $734M

Votorantim raises $570M, renamed Nexa Resources

Nyrstar raises $116M and halts mine disposals

MAG Silver raises $48M

Sprott funds Century Tailings $45M debt

Central Asia Metals buys Lynx Resources (Sasa mine) for $403M

China NFC to buy 41% of PT Dairi for $198M

Bank of China agrees $120M loan (within EBRD loan of $350M) for Shalkiya project

New mine supply (kt/a Zn)

Castellanos	75kt	Q4 2017
Dugald River	170kt	Q4 2017
Century Tailings	270kt	2018
Gamsberg	250kt	2018
Lady Lorretta (Mt Isa)	160kt	2018
Empire State	55kt	2018
Shalkiya	125kt	2019
Rey de Plata	40kt	2019
Woodlawn	40kt	2019
Zhairem	160kt	2020

New mine supply will create a balanced concentrate market in 2018, a surplus in 2019 and significant surplus from 2020

Probable projects are the most likely candidates for new mines

The absence of traditional bank and equity finance is hindering construction of the larger zinc mine projects

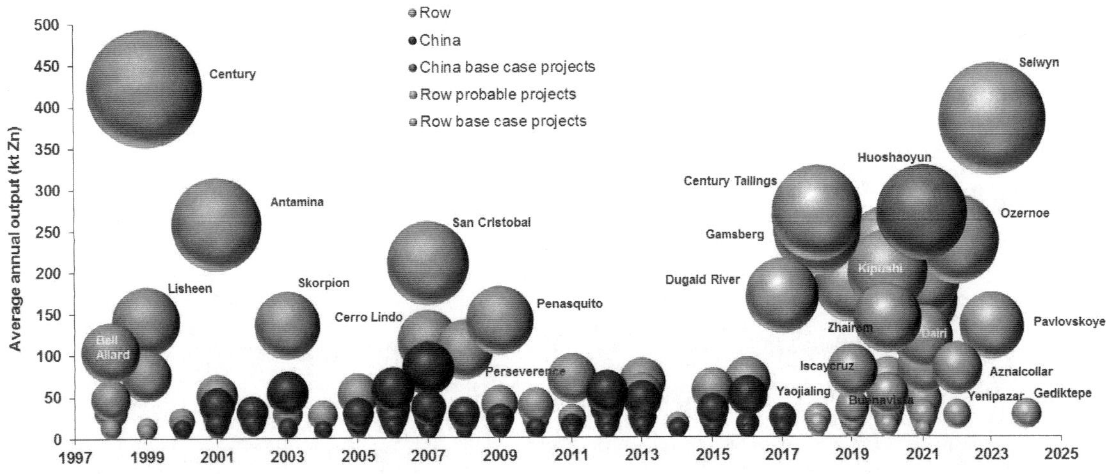

Source: Wood Mackenzie

Once permitted and financed zinc mines can be built quickly

Price expectation is the main driver

Typical construction time has been less than three years

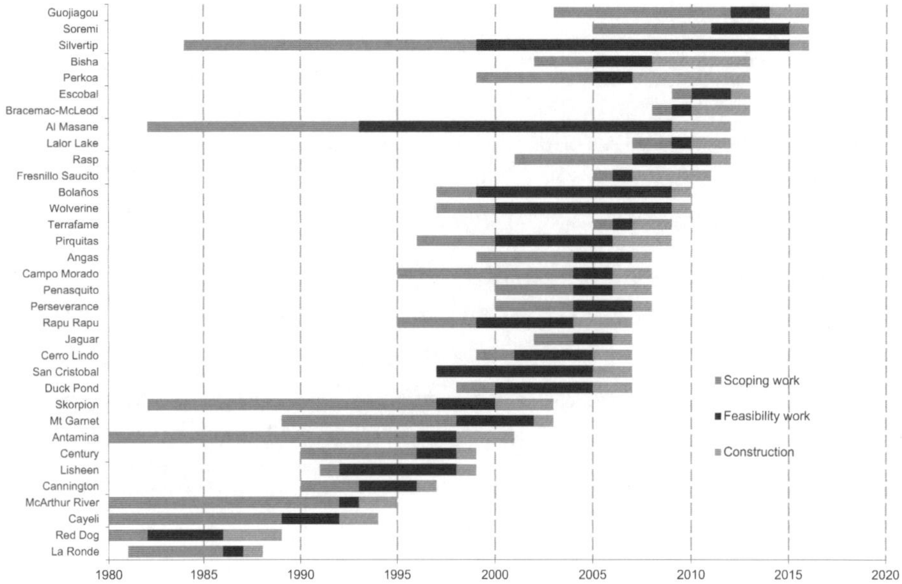

Source: Wood Mackenzie

Chinese environmental shutdowns are a potential downside risk to concentrate supply

woodmac.com 17

Exchange stocks fall to 6 days of global consumption

Zinc price volatility increases once stocks fall below the equivalent of 10 days of global consumption. This was breached in June 2017 and continued to fall from 14 days in January 2017 to 6 days in January 2018

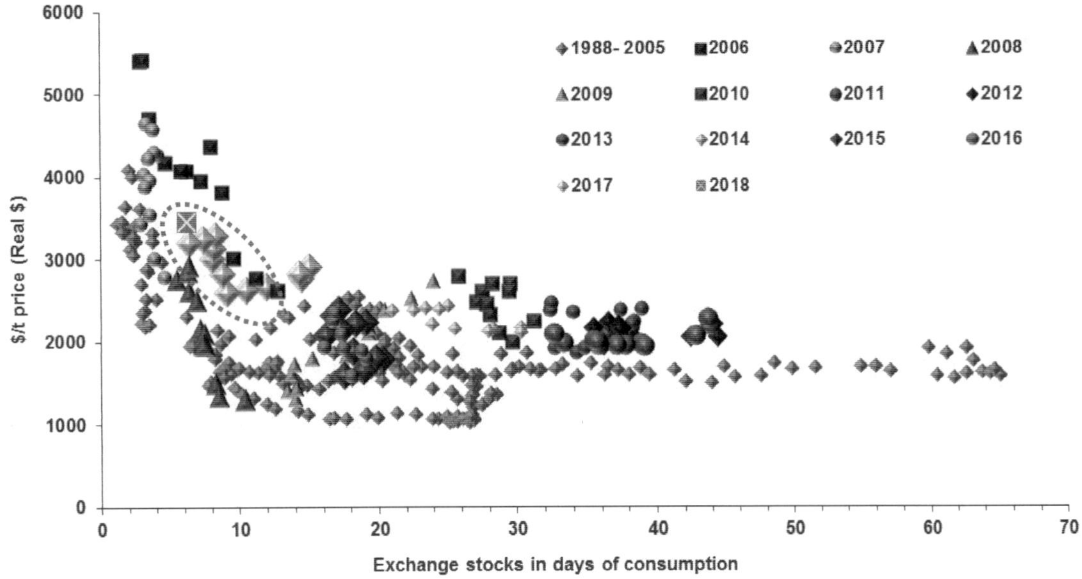

Source: Wood Mackenzie

woodmac.com 18

Low LME stocks have led to increased backwardations

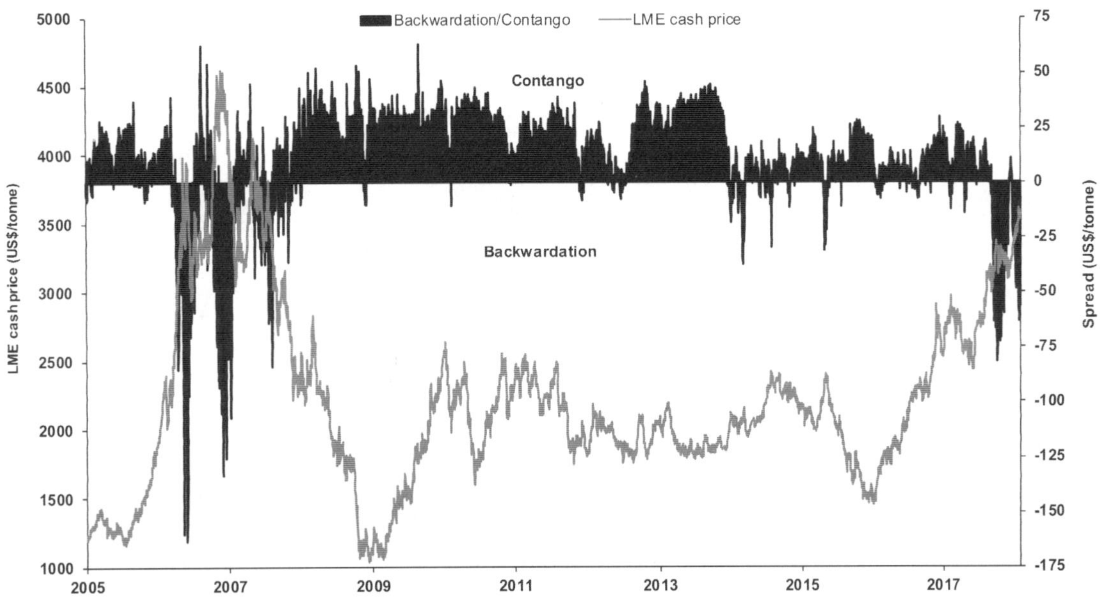

Source: LME, Wood Mackenzie

Increasing tightness in the refined market will provide fundamental support for higher prices

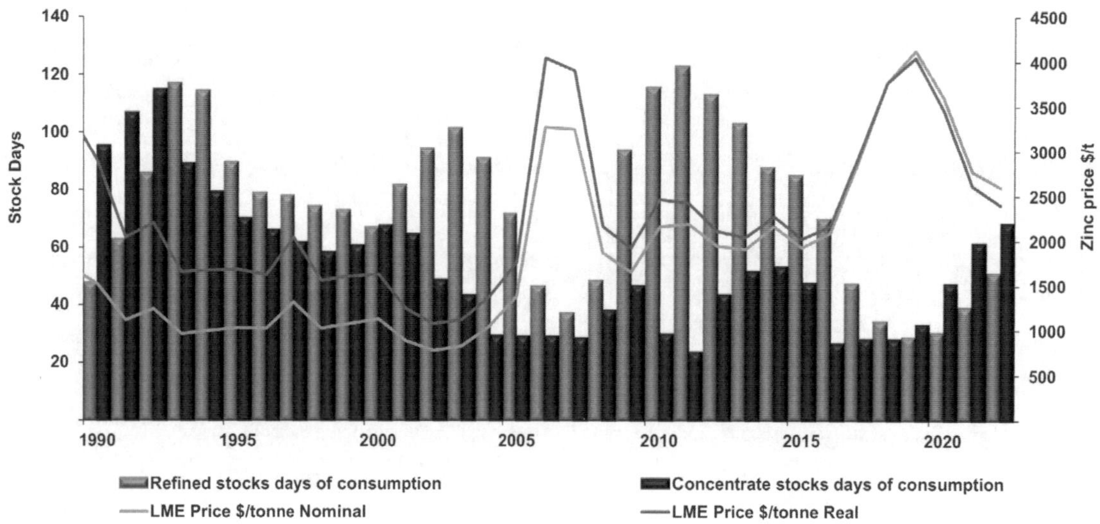

Source: Wood Mackenzie

High prices increase the risk of demand deferment and destruction

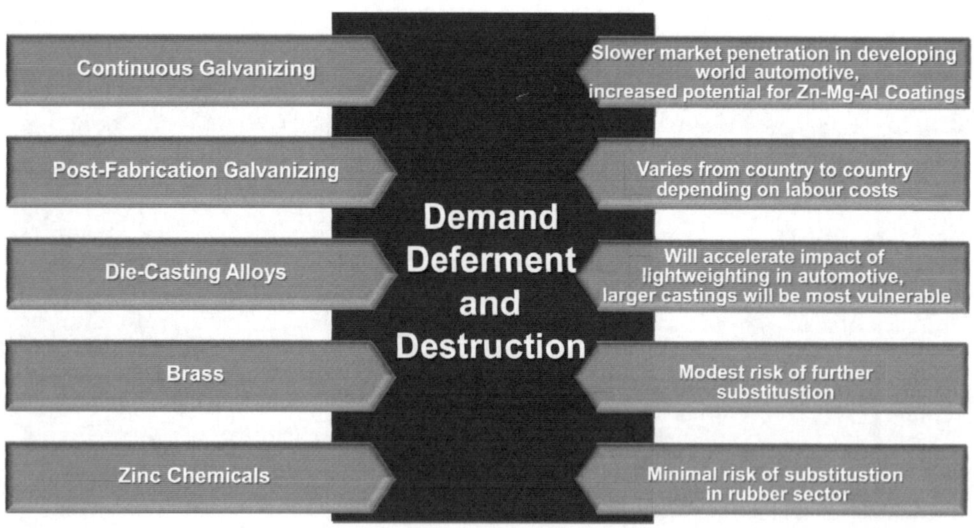

Summary

Mine closures and production cuts have reduced concentrate stocks to record lows, disrupting concentrate flows, forcing treatment charges lower, and constrained growth in refined production

Refined metal stocks are falling to historically low levels, driving the zinc price to a cyclical peak in 2019

The key risks for 2018 are on the upside:

- Environmental regs. in China constrain mine production growth
- Mine projects do not deliver planned tonnage on time

Supply constraints for mines and smelters mean that zinc prices will remain at historically high levels. However, risk that demand will be deferred or be destroyed is growing

Zinc and lead research team

Zinc Markets

Andrew Thomas
Principal Analyst - Demand
London, UK

Chris Parker
Research Director
London, UK

Jonathan Leng
Principal Analyst - Supply
London, UK

Sherry Shen
Consultant, Zinc & Lead
Beijing, China

Lead and Zinc Mine Costs

Georgina Marchorst
Manager, Lead & Zinc Mine Costs
London, UK

Joan Arratia
Cost Analyst, Lead & Zinc
London, UK

Ruben Arratia
Research Director, Base & Precious Metals
Lima, Peru

Maria Rosa Gobitz
Associate – Base & Precious Metals
New York, USA

Sifang Liu
Senior Consultant – Metals Cost
Beijing, China

Hun Hao Chuah
Analyst – Metals Research
Singapore

Lead Markets

Farid Ahmed
Principal Analyst - Demand
London, UK

Giles Lloyd
Principal Analyst - Supply
London, UK

Zinc Smelter Costs

John Anderson
Principal Analyst Zinc Smelters
London, UK

Jonathan Leng

Principal Analyst Zinc Markets

Biography

Jonathan is a principal analyst focused on zinc mine and smelter supply, plus concentrate analysis. Prior to his current role, Jonathan was manager of the zinc, lead and gold mine cost research teams. He has 20 years of metals and mining experience, primarily in base metal, ferroalloy and gold research.

Prior to joining Wood Mackenzie, he worked as a geologist for Croesus Gold in Australia and metals research at CRU International. Jonathan also worked in construction cost management at Davis Langdon.

Jonathan holds a M.Sc. Mining and Mineral Resource Evaluation, B.Sc. (hons) Geology, and PGDip in Construction Cost Management.

Jonathan.Leng@woodmac.com

+44 1483 596 626

Wood Mackenzie
A Verisk Business

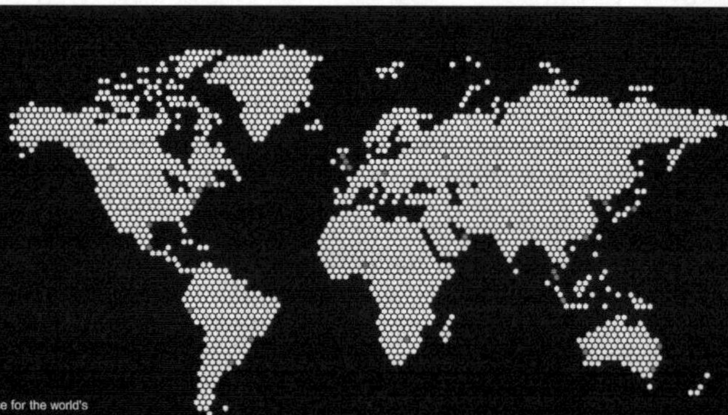

Europe	+44 131 243 4400
Americas	+1 713 470 1600
Asia Pacific	+65 6518 0800
Email	contactus@woodmac.com
Website	www.woodmac.com

Wood Mackenzie™, a Verisk business, is a trusted source of commercial intelligence for the world's natural resources sector. We empower clients to make better strategic decisions, providing objective analysis and advice on assets, companies and markets. **For more information visit: www.woodmac.com**

Macquarie Research | Commodities

MACQUARIE

Zinc Outlook

Now what?

Macquarie Commodities Research

Macquarie Capital (Europe) Ltd

28 Ropemaker Street

London EC2Y 9HD, UK

Vivienne Lloyd

+44 20 3037 4530
vivienne.lloyd@macquarie.com

February 2018

Please read Disclaimer on Pages 23-25

Macquarie Research | Commodities

MACQUARIE

Zinc outlook – now what?

- ➔ Global macro picture
- ➔ China
- ➔ Zinc

LME zinc prices ($/t)

Macquarie Research | Commodities

MACQUARIE

Zinc outlook – now what?

- → Global macro picture
- → China
- → Zinc

Source: Bloomberg, February 2018

Page 3

Macquarie Research | Commodities

MACQUARIE

2017 good for most commodities, especially metals

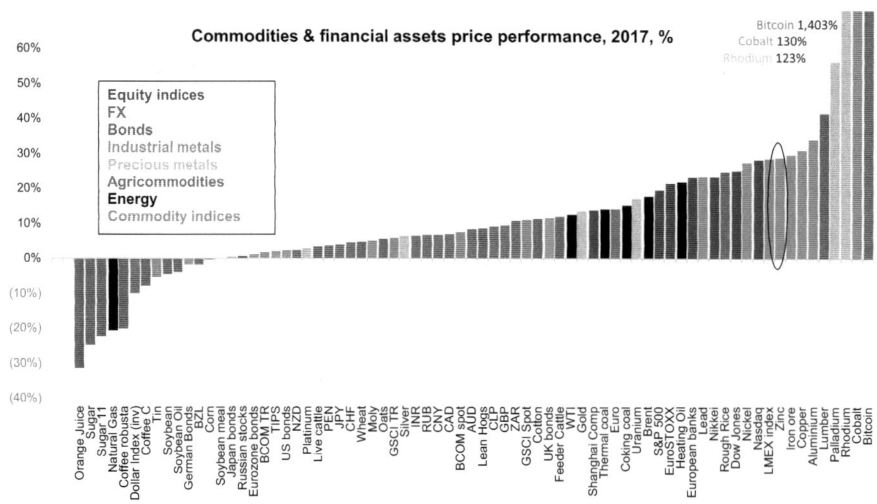

Source: Bloomberg, Macquarie Research, February 2018

Page 4

Macquarie Research | Commodities

MACQUARIE

In general, medium term we like precious, are cautious base, bearish bulks, and we would short EV battery metals

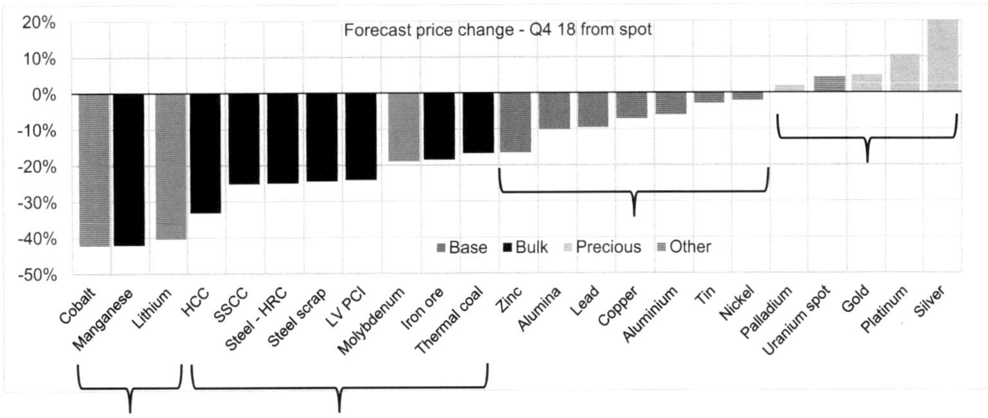

Source: Macquarie Research, February 2018

Page 5

Macquarie Research | Commodities

MACQUARIE

Sustained synchronised global growth has underpinned wider commodities recovery

Global industrial production, 3m/3m, % annualised

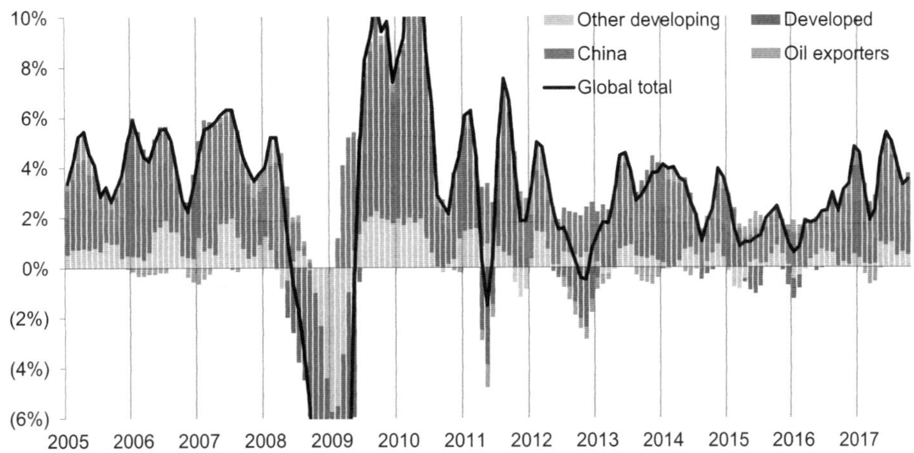

Source: National Sources, Macquarie Research, February 2018

Page 6

2018 will be another good year for global GDP

- Global GDP growth has accelerated to its best since 2011, and while growth rates are likely to "top-out" in mid-year 2018, full-year growth will be slightly higher than in 2017 on strong US and Eurozone performances. The long shadow of the GFC is abating.
- China's economy will remain by far the fastest-growing but will slow from 6.9% in 2017 to the mid-6%s.

Global GDP growth, % YoY

GDP growth by region/country, % per year

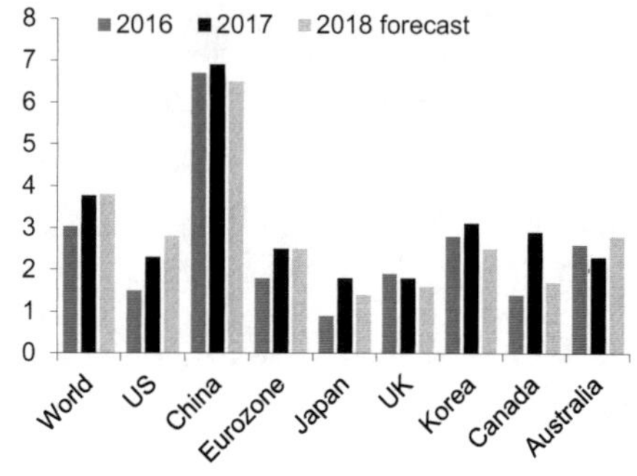

Source: Macrobond, Macquarie Research, February 2018

Macquarie Research | Commodities
MACQUARIE

Equity sell-off …event catalysts? 1. report in US of housing income inflation; 2. USD bounce.

Commodity price + rate indices (12mth rolling)

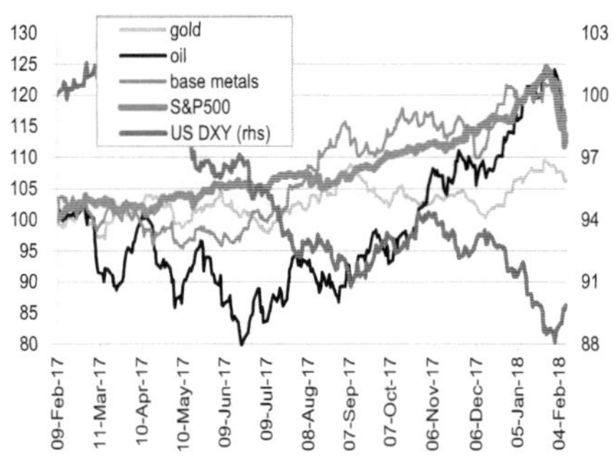

Resources sector indices (12mth rolling)

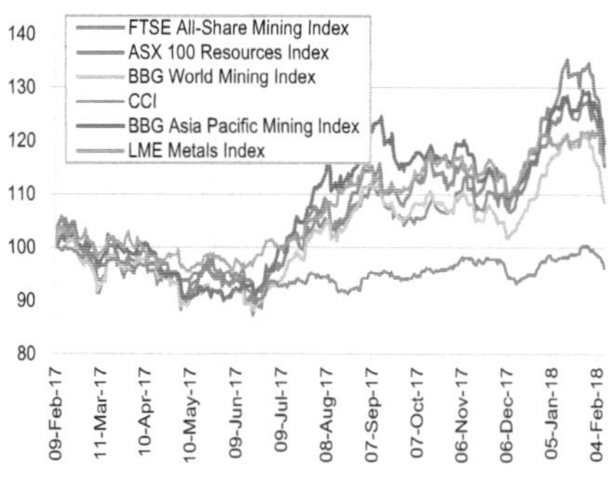

Source: Bloomberg, February 2018

Macquarie Research | Commodities

MACQUARIE

Zinc equities vs. metal price

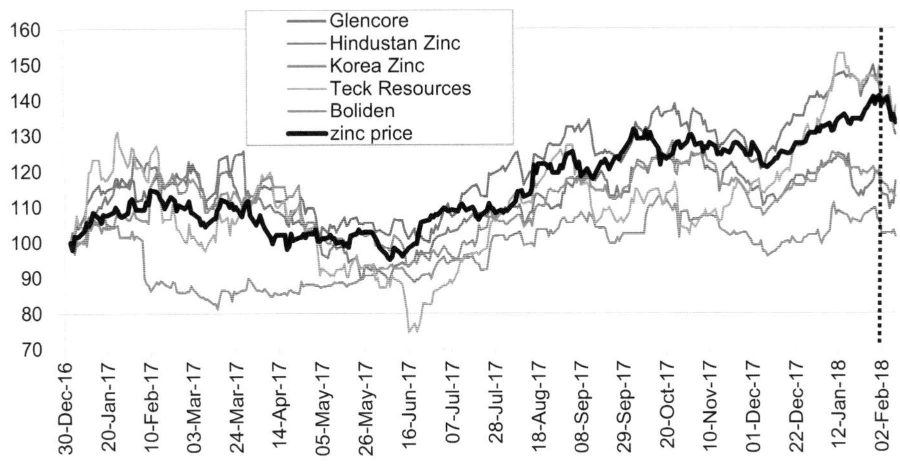

Source: Bloomberg, February 2018

Page 9

Macquarie Research | Commodities

MACQUARIE

Zinc outlook – now what?

- ➔ Global macro picture
- ➔ China
- ➔ Zinc

Source: Bloomberg, February 2018

Page 10

China: slowing and deleveraging expected

- The biggest risk is China. Not so much for the global economy – China has a large but not dominant weight there – but in commodities, where it still accounts for ~50% of demand for most.
- China's latest round of macro data gave a mixed picture – with property and external demand strong but imports, credit and FAI less robust. Our China economist warns that 2018 could be a year of deleveraging in the real economy.

China remains dominant in hard commodities

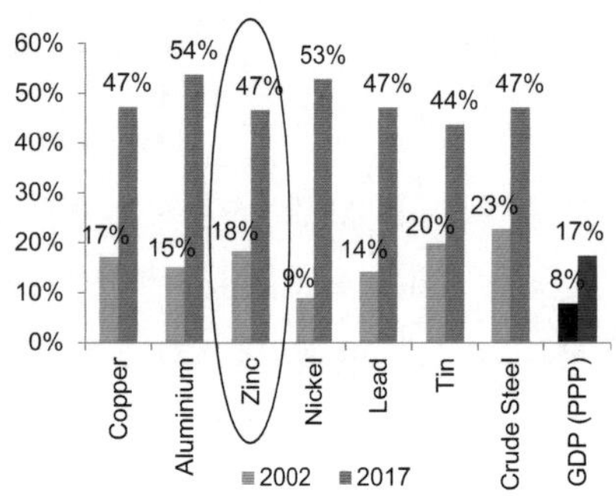

Source: National data, Macquarie Research, February 2018

IP remains strong but FAI is slowing, while the property market boom is over, but no signs yet of bust (all %

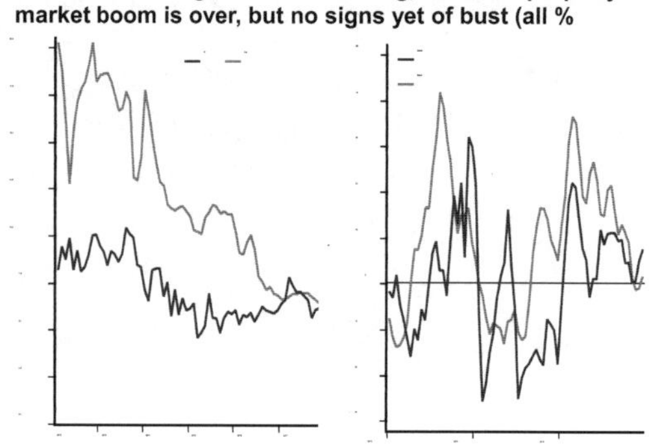

Macquarie Research | Commodities

MACQUARIE

Per capita consumption is beginning to mature, particularly for crude steel

Crude Steel consumption

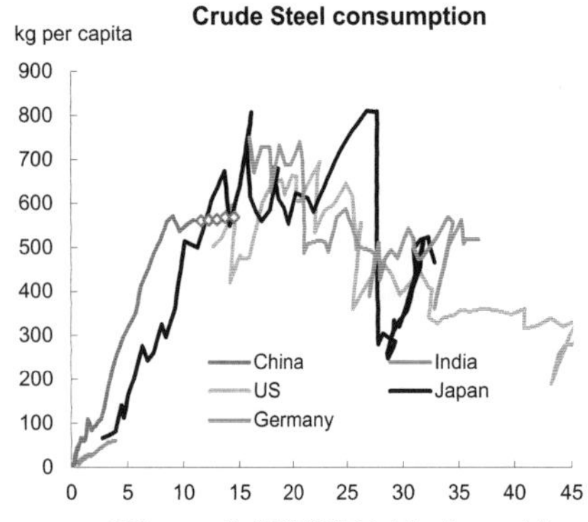

Zinc consumption

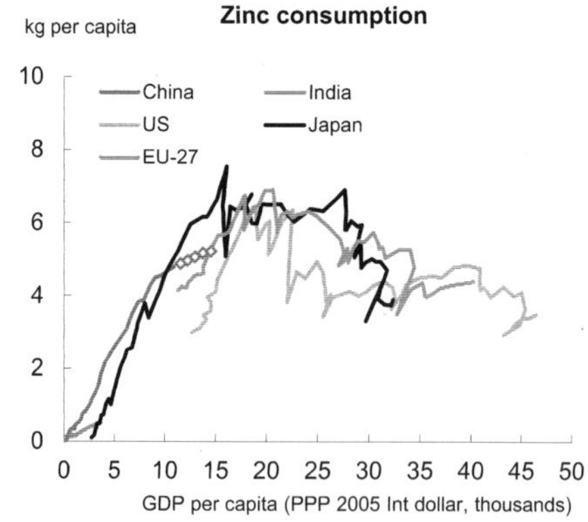

Source: International Study Groups, CRU, Wood Mac, SMM, Antaike, worldsteel, Macquarie Research, February 2018

Macquarie Research | Commodities

MACQUARIE

Zinc outlook – now what?

- ➔ Global macro picture
- ➔ China
- ➔ Zinc

Source: Bloomberg, February 2018

Page 13

Macquarie Research | Commodities

MACQUARIE

Zinc: The story that played out

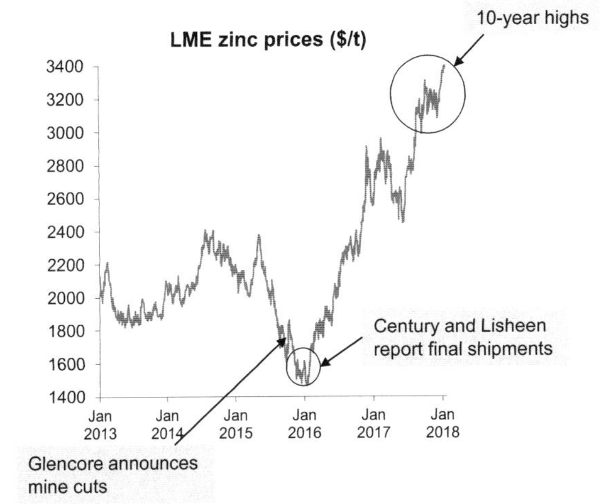

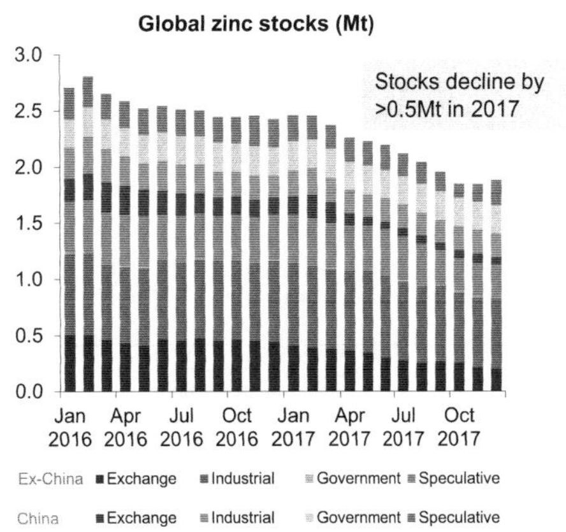

Source: LME, Macquarie Research, February 2018

Page 14

Macquarie Research | Commodities

MACQUARIE

Ingot imports rally strongly as smelter growth struggles from low raw material availability

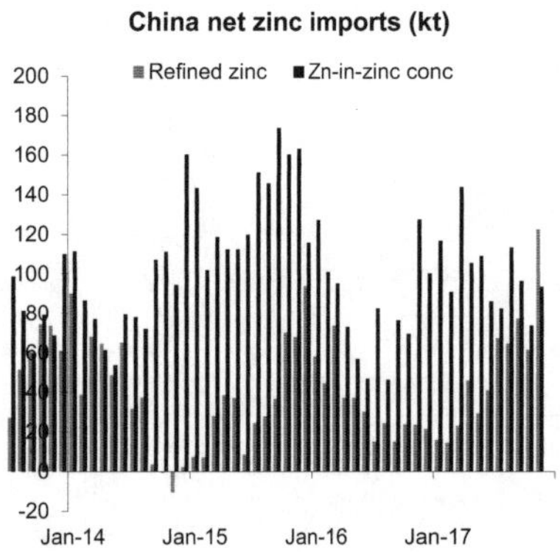

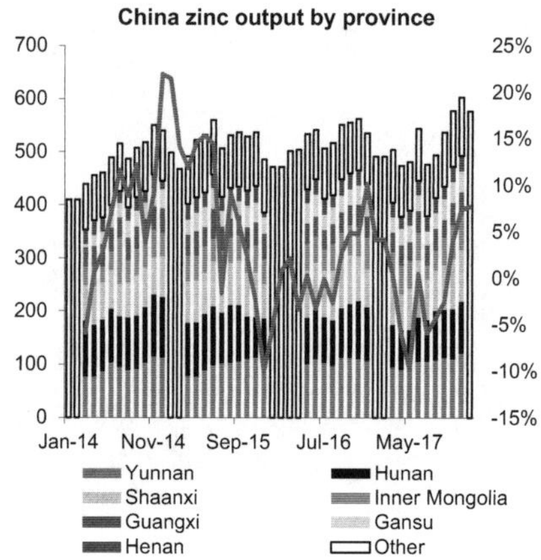

Source: China Customs, NBS, Macquarie Research, February 2018

Page 15

Macquarie Research | Commodities

MACQUARIE

Import origin is interesting – Spanish slab appears and Peruvian concs shipments have recovered

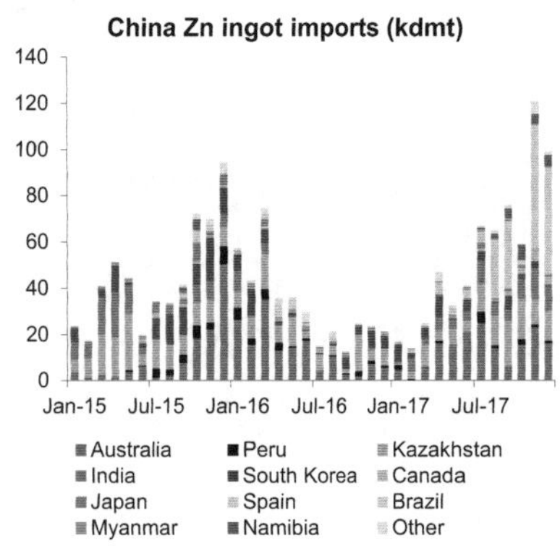

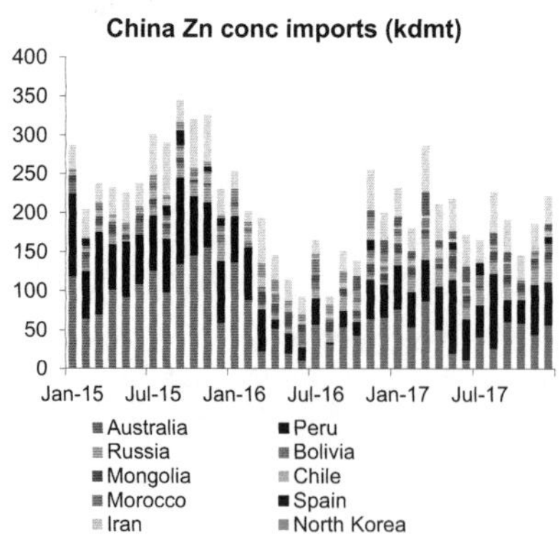

Source: China Customs, NBS, Macquarie Research, February 2018

Page 16

Spot terms stabilising, still no ingot premium lift-off as spreads staying tight on warrant availability

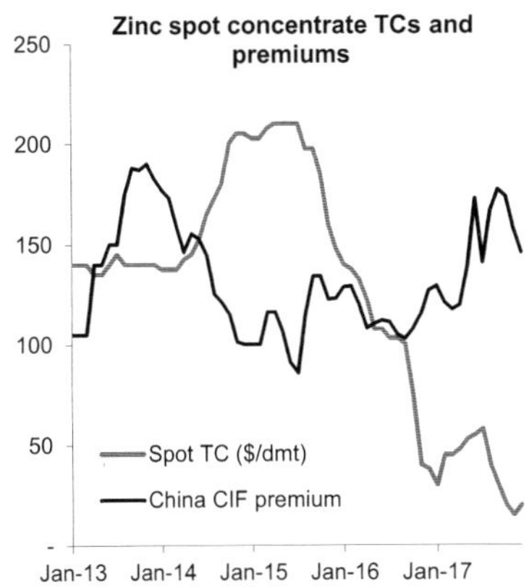

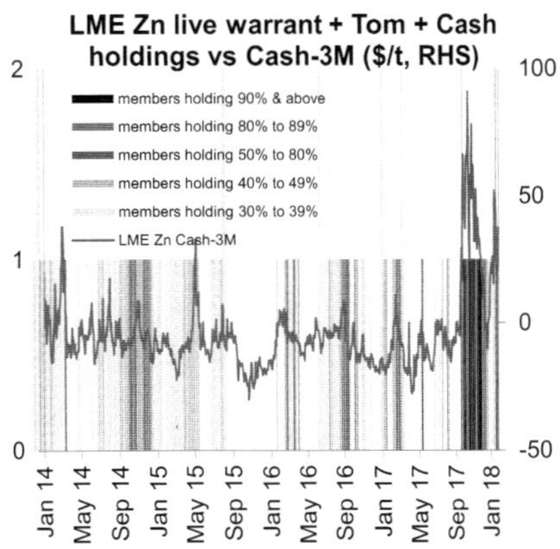

Source: CRU, Metal Bulletin, LME, SHFE, Macquarie Research, February 2018

Near-term upside risk on post-winter bounce-back, and DCA output holding up despite Zn:Al ratio

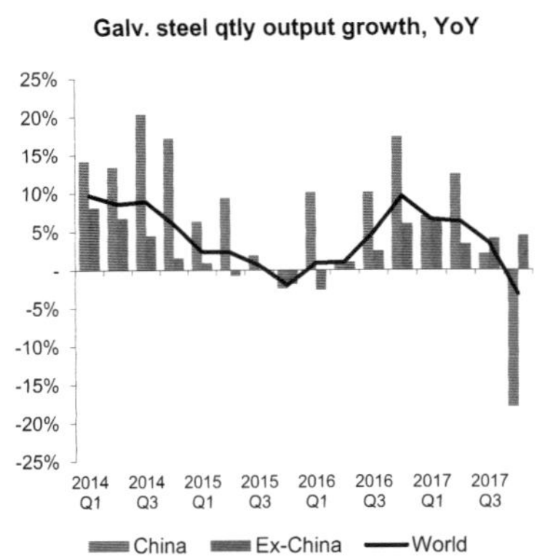

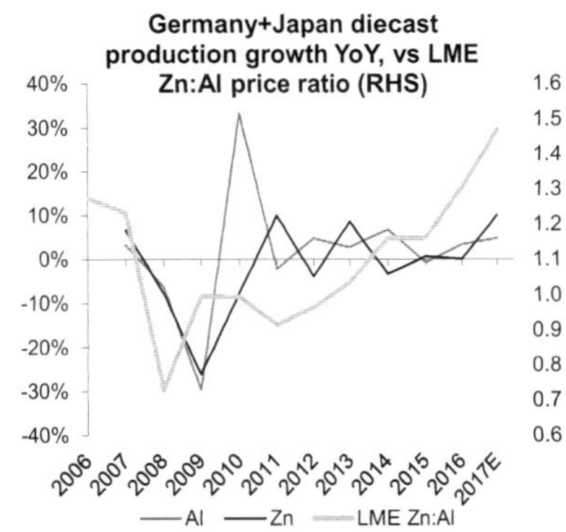

Source: CRU, VDD, Japan Diecasters Association, LME, Macquarie Research, February 2018

Macquarie Research | Commodities

MACQUARIE

But high prices have begun to attract new mine supply, returning balance to the market

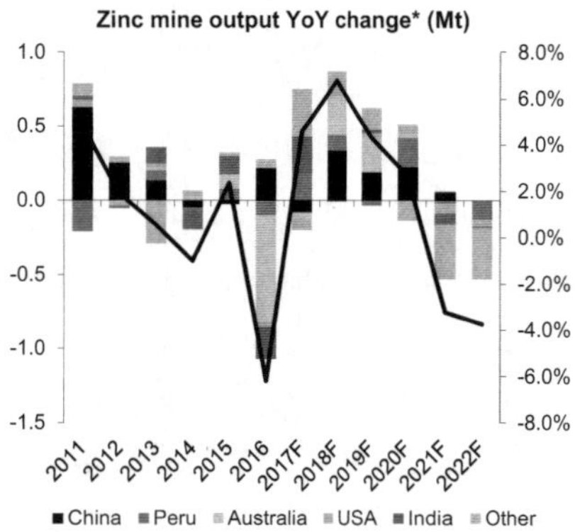

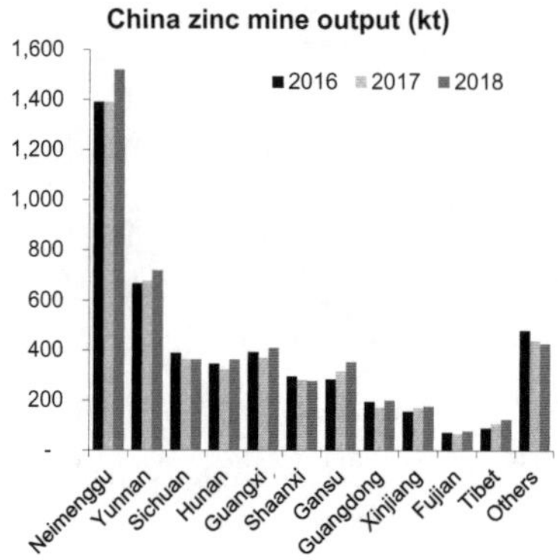

Source: Company reports, ILZSG, LME, CRU, Wood Mackenzie, Macquarie Research, February 2018

Page 19

Macquarie Research | Commodities

MACQUARIE

…and a slowdown in Chinese property will soften ferrous commodities, and likely hit zinc

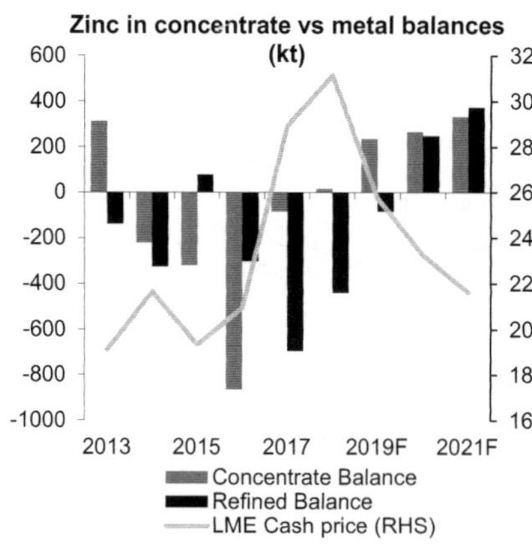

Source: Company reports, ILZSG, LME, CRU, Wood Mackenzie, Macquarie Research, February 2018

Page 20

Macquarie Research | Commodities

MACQUARIE

Zinc supply-demand balance

'000t zinc	2014	2015	2016	2017	2018F	2019F	2020F	2021F	2022F
Mine production	12,635	12,931	12,130	12,699	13,237	14,003	14,633	15,102	15,416
% Change YoY	-1.0%	2.3%	-6.2%	4.7%	4.2%	5.8%	4.5%	3.2%	2.1%
Concs Balance	**-219**	**-319**	**-865**	**-62**	**36**	**257**	**288**	**354**	**419**
Refined production	13,203	13,636	13,466	13,353	13,837	14,432	15,082	15,534	15,845
% Change YoY	2.9%	3.3%	-1.2%	-0.8%	3.6%	4.3%	4.5%	3.0%	2.0%
Consumption	13,528	13,560	13,767	14,067	14,296	14,536	14,855	15,183	15,520
% Change YoY	4.3%	0.2%	1.5%	2.2%	1.6%	1.7%	2.2%	2.2%	2.2%
Refined Balance	**-325**	**75**	**-301**	**-715**	**-459**	**-104**	**226**	**351**	**325**
Estimated total stocks	2698	2748	2427	1884	1425	1321	1547	1899	2223
Weeks of consumption	10.4	10.5	9.2	7.0	5.2	4.7	5.4	6.5	7.4
LME Cash Price ($/t)	**2,164**	**1,932**	**2,092**	**2,894**	**3,113**	**2,575**	**2,325**	**2,163**	**2,175**
LME Cash Price (c/lb)	98	88	95	131	141	117	105	98	99

Source: ILZSG, CRU, Wood Mackenzie, LME, Macquarie Research, February 2018

Macquarie Research | Commodities

MACQUARIE

Thank you

Vivienne Lloyd
+44 20 3037 4530
vivienne.lloyd@macquarie.com

Important disclosures:

Recommendation definitions

Macquarie - Australia/New Zealand

Outperform – return > 3% in excess of benchmark return
Neutral – return within 3% of benchmark return
Underperform – return > 3% below benchmark return

Benchmark return is determined by long term nominal GDP growth plus 12 month forward market dividend yield.

Macquarie – Asia/Europe

Outperform – expected return >+10%
Neutral – expected return from -10% to +10%
Underperform – expected <-10%

Macquarie - South Africa

Outperform – return > 10% in excess of benchmark return
Neutral – return within 10% of benchmark return
Underperform – return > 10% below benchmark return

Macquarie - Canada

Outperform – return > 5% in excess of benchmark return
Neutral – return within 5% of benchmark return
Underperform – return > 5% below benchmark return

Macquarie - USA

Outperform – return > 5% in excess of benchmark return
Neutral – return within 5% of benchmark return
Underperform – return > 5% below benchmark return

Volatility index definition*

This is calculated from the volatility of historic price movements.

Very high–highest risk – Stock should be expected to move up or down 60-100% in a year – investors should be aware this stock is highly speculative.

High – stock should be expected to move up or down at least 40-60% in a year – investors should be aware this stock could be speculative.

Medium – stock should be expected to move up or down at least 30-40% in a year.

Low–medium – stock should be expected to move up or down at least 25-30% in a year.

Low – stock should be expected to move up or down at least 15-25% in a year.

* Applicable to Australian/NZ stocks only

Recommendation – 12 months

Note: Quant recommendations may differ from Fundamental Analyst recommendations

Financial definitions

All "Adjusted" data items have had the following adjustments made:

Added back: goodwill amortisation, provision for catastrophe reserves, IFRS derivatives & hedging, IFRS impairments & IFRS interest expense
Excluded: non recurring items, asset revals, property revals, appraisal value uplift, preference dividends & minority interests

EPS = adjusted net profit /efpowa*
ROA = adjusted ebit / average total assets
ROA Banks/Insurance = adjusted net profit /average total assets
ROE = adjusted net profit / average shareholders funds
Gross cashflow = adjusted net profit + depreciation
*equivalent fully paid ordinary weighted average number of shares

All Reported numbers for Australian/NZ listed stocks are modelled under IFRS (International Financial Reporting Standards).

Recommendation proportions – For quarter ending 31 December 2017

	AU/NZ	Asia	RSA	USA	CA	EUR	
Outperform	51.82%	55.57%	44.05%	45.06%	60.00%	42.51%	(for global coverage by Macquarie, 4.36% of stocks followed are investment banking clients)
Neutral	35.40%	28.60%	36.90%	47.59%	28.67%	40.42%	(for global coverage by Macquarie, 2.58% of stocks followed are investment banking clients)
Underperform	12.77%	15.83%	19.05%	7.34%	11.33%	17.07%	(for global coverage by Macquarie, 0.69% of stocks followed are investment banking clients)

Company-Specific Disclosures:

Important disclosure information regarding the subject companies covered in this report is available at www.macquarie.com/research/disclosures

Analyst Certification:

We hereby certify that all of the views expressed in this report accurately reflect our personal views about the subject company or companies and its or their securities. We also certify that no part of our compensation was, is or will be, directly or indirectly, related to the specific recommendations or views expressed in this report. The Analysts responsible for preparing this report receive compensation from Macquarie that is based upon various factors including Macquarie Group Ltd total revenues, a portion of which are generated by Macquarie Group's Investment Banking activities.

General Disclaimers:

Macquarie Securities (Australia) Ltd; Macquarie Capital (Europe) Ltd; Macquarie Capital Markets Canada Ltd; Macquarie Capital Markets North America Ltd; Macquarie Capital (USA) Inc; Macquarie Capital Limited and Macquarie Capital Limited, Taiwan Securities Branch; Macquarie Capital Securities (Singapore) Pte Ltd; Macquarie Securities (NZ) Ltd; Macquarie Equities South Africa (Pty) Ltd.; Macquarie Capital Securities (India) Pvt Ltd; Macquarie Capital Securities (Malaysia) Sdn Bhd; Macquarie Securities Korea Limited and Macquarie Securities (Thailand) Ltd are not authorized deposit-taking institutions for the purposes of the Banking Act 1959 (Commonwealth of Australia), and their obligations do not represent deposits or other liabilities of Macquarie Bank Limited ABN 46 008 583 542 (MBL) or MGL. MBL does not guarantee or otherwise provide assurance in respect of the obligations of any of the above mentioned entities. MGL provides a guarantee to the Monetary Authority of Singapore in respect of the obligations and liabilities of Macquarie Capital Securities (Singapore) Pte Ltd for up to SGD 35 million. This research has been prepared for the general use of the wholesale clients of the Macquarie Group and must not be copied, either in whole or in part, or distributed to any other person. If you are not the intended recipient you must not use or disclose the information in this research in any way. If you received it in error, please tell us immediately by return e-mail and delete the document. We do not guarantee the integrity of any e-mails or attached files and are not responsible for any changes made to them by any other person. MGL has established and implemented a conflicts policy at group level (which may be revised and updated from time to time) (the "Conflicts Policy") pursuant to regulatory requirements (including the FCA Rules) which sets out how we must seek to identify and manage all material conflicts of interest. Nothing in this research shall be construed as a solicitation to buy or sell any security or product, or to engage in or refrain from engaging in any transaction. In preparing this research, we did not take into account your investment objectives, financial situation or particular needs. Macquarie salespeople, traders and other professionals may provide oral or written market commentary or trading strategies to our clients that reflect opinions which are contrary to the opinions expressed in this research. Macquarie Research produces a variety of research products including, but not limited to, fundamental analysis, macro-economic analysis, quantitative analysis, and trade ideas. Recommendations contained in one type of research product may differ from recommendations contained in other types of research, whether as a result of differing time horizons, methodologies, or otherwise. Before making an investment decision on the basis of this research, you need to consider, with or without the assistance of an adviser, whether the advice is appropriate in light of your particular investment needs, objectives and financial circumstances. There are risks involved in securities trading. The price of securities can and does fluctuate, and an individual security may even become valueless. International investors are reminded of the additional risks inherent in international investments, such as currency fluctuations and international stock market or economic conditions, which may adversely affect the value of the investment. This research is based on information obtained from sources believed to be reliable but we do not make any representation or warranty that it is accurate, complete or up to date. We accept no obligation to correct or update the information or opinions in it. Opinions expressed are subject to change without notice. No member of the Macquarie Group accepts any liability whatsoever for any direct, indirect, consequential or other loss arising from any use of this research and/or further communication in relation to this research. Clients should contact analysts at, and execute transactions through, a Macquarie Group entity in their home jurisdiction unless governing law permits otherwise. The date and timestamp for above share price and market cap is the closed price of the price date. #CLOSE is the final price at which the security is traded in the relevant exchange on the date indicated.

Macquarie Research | Commodities

MACQUARIE

Contact Macquarie Capital (USA) Inc.

SALES

US

Khristina McLaughlin (212) 231 8012
Michael Burguieres (212) 231 2506
Michael Cline (212) 231 2524
Anthony Espinosa (212) 231 0937
Robert McKeon (212) 231 2573
Rob Moderelli (212) 231 2495
Heidi Muccifori (212) 231 2466
Tad Nacheff (212) 231 2615
Mark Orsi (415) 762 5002
Shaival Patel (212) 231 8002
Jason Philosophos (212) 231 2567
Troy Pilalas (415) 762 5025
Jason Rubin (212) 231 8011
Ryan Scott (212) 231 2505
Alex Shaner (617) 598 2503
Denver Smith (617) 598 2507
Lisa Sowell (212) 231 1787

Email Addresses
firstname.surname@macquarie.com

Australia/NZ

Leighton Patrick (212) 231 2552
Michael McNair (212) 231 2571
Jay Shyam (212) 231 2491

Asia

Jeff Evans (212) 231 2559
Paul Colaco (415) 762 5003
Aya Hirose (212) 231 2436
Shunsuke Kuriyama (212) 231 2562
Mark Lawrence (212) 231 2516
Todd Narter (415) 762 5005
Brad Neill (212) 231 0973
Jeff Seo (212) 231 2422
Jean Zhang (212) 231 6397

Europe

Chris Carr (212) 231 6398
Juliette Lafille-Mowatt (212) 231 0977
Doug Stone (212) 231 2606

SALES TRADING

US

J.T. Cacciabaudo (212) 231 6381
Chris Baker (212) 231 8064
Steven Belge (617) 598 2515
Greg Calvino (212) 231 0902
Frank Casey (617) 598 2509
Adam Cavise (212) 231 8043
Mike Gray (212) 231 2555
Brad Hoenig (212) 231 8089
Carlos Jimenez (212) 231 0883
Allen Jordan (212) 231 0947
Marty Livingston (212) 231 2523
Matt McCloskey (212) 231 8024
Rob Salmon (415) 762 5009
Patrick Scanlan (312) 660 9025
Tarik Turner (212) 231 6365

International

Robert Lenihan (212) 231 8030
Chris Reale (212) 231 2555
Madeline Richardson (212) 231 2487

TRADING

Brian Ross (212) 231 2534
Ed Bartleson (212) 231 2555
Joe Daly (212) 231 2542
Pete McQuaid (212) 231 2459
Michael Nevitt (212) 231 2408
Chris Zarins (212) 231 0882

EXECUTION SERVICES (PT)

Matt McNamara (212) 231 0900
Jesse Busch (212) 231 0900
Jessica Pepin (212) 231 0900
Ben Salup (212) 231 0900

EXECUTION SERVICES (EE)

Stuart McWilliam (212) 231 0925
Matt Hor (212) 231 0300
Marc Rosa (212) 231 0300

CORPORATE ACCESS

Margaret Shergalis (212) 231 1348
Karen Bader (212) 231 6364
Katie Gray (212) 231 1318
Angie Mouta (212) 231 1823
Shelley O'Donnell (212) 231 1798

Macquarie Capital (USA) Inc.
125 W. 55th Street, Level 22
New York, NY 10019
(212) 231 1000

Macquarie Capital (USA) Inc.
2 Embarcadero Center, Suite 200
San Francisco, CA 94111
(415) 762 5000

Macquarie Capital (USA) Inc.
225 Franklin Street, 17th Floor
Boston, MA 02110
(617) 598 2500

Macquarie Capital (USA) Inc.
1 North Wacker Drive, Level 29
Chicago, IL 60606
(312) 660 9215

Macquarie Capital (USA) Inc
1225 17th Street, Suite 1600
Denver, CO, 80202
(303) 952 2800

24 Hour Toll-Free
1-888 MAC STOCK
1-888 622 7862

CRU

Zinc Market Outlook

Will 2018 H1 mark the turning point?

International Zinc & Oxide Conferences, Carlsbad, USA

11-14 February 2018

CRU THE INDEPENDENT AUTHORITY MINING I METALS I FERTILIZERS

1

How much higher can prices go?

- How tight is the market now, following last year's record metal deficits?
- How strong will the supply-side response be?
- How are high prices affecting demand?
- What does this all mean for the market balance and prices going forward?

www.crugroup.com

CRU THE INDEPENDENT AUTHORITY
MINING | METALS | FERTILIZERS

Concentrate stocks were exhausted last year

cumulative zinc market surplus; adjusted; million tonnes

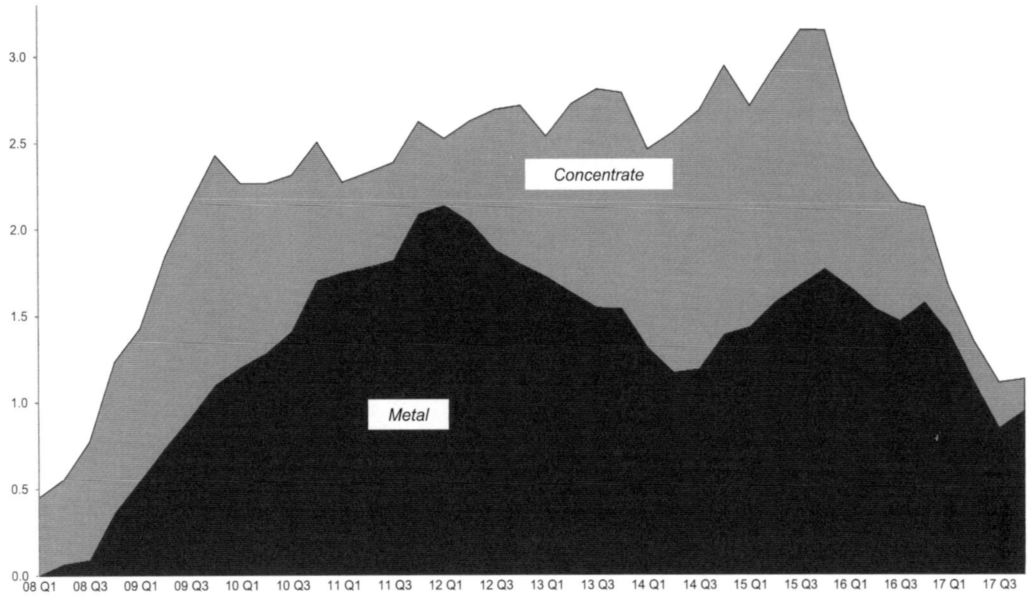

Source: January 2018 Zinc Market Outlook www.crugroup.com

CRU THE INDEPENDENT AUTHORITY
MINING | METALS | FERTILIZERS

The positive SHFE/LME arb led to a surge in China's net imports of refined metal

The arb became more negative for imports in January 2018
SHFE/LME arb, $/t

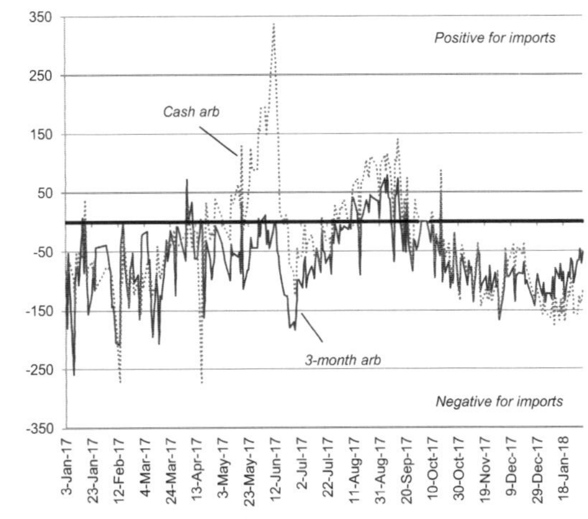

Data: LME, SHFE, CRU

China's net imports surged in November and December 2017
Chinese refined imports (exports), '000t

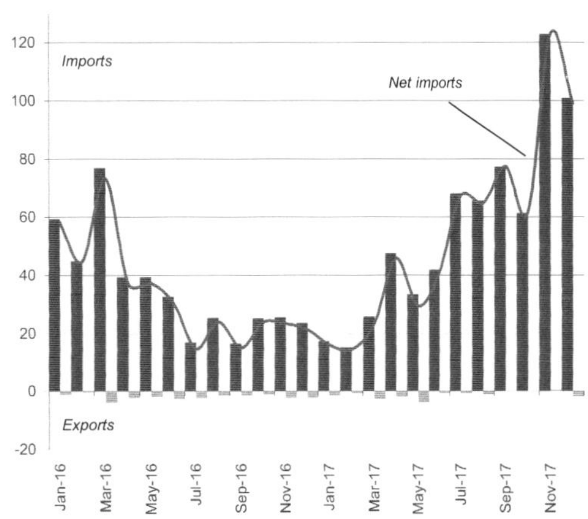

Data: GTIS, CRU

Source: January 2018 Zinc Market Outlook www.crugroup.com

2018 contract premia in the USA increased by the most

Spot premia

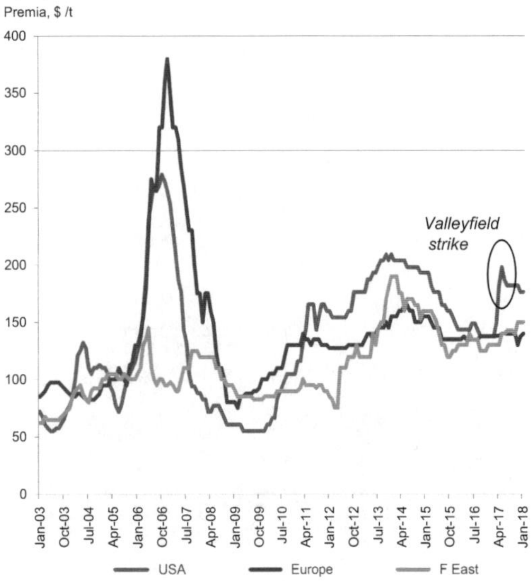

Data: CRU

Long-term contract premia

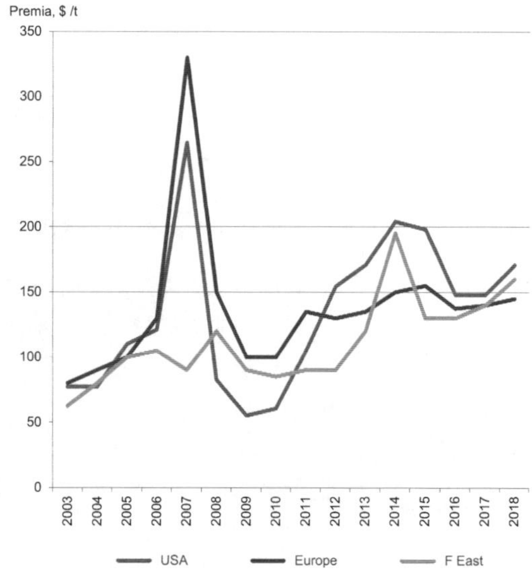

Data: CRU

Source: CRU January 2018 www.crugroup.com

Positive investment flows supported zinc's price

US$ per tonne

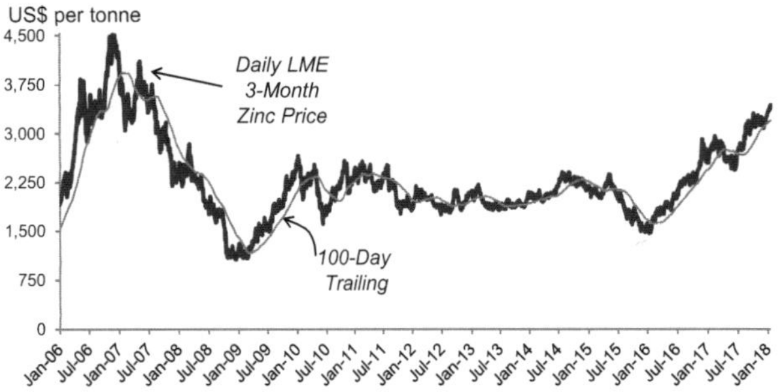

LME cash to 3-month contango (US$/t)

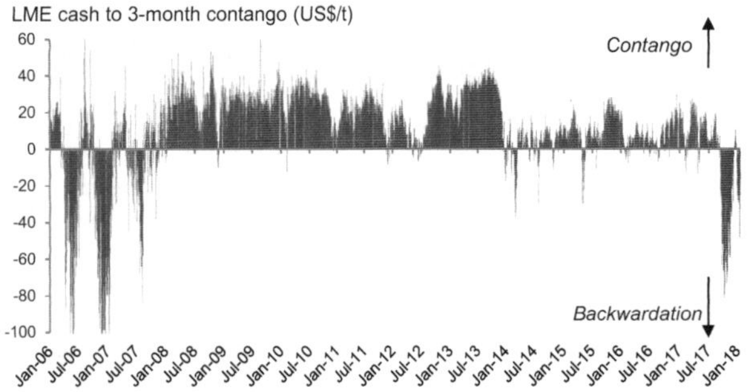

net forward position, '000t zinc

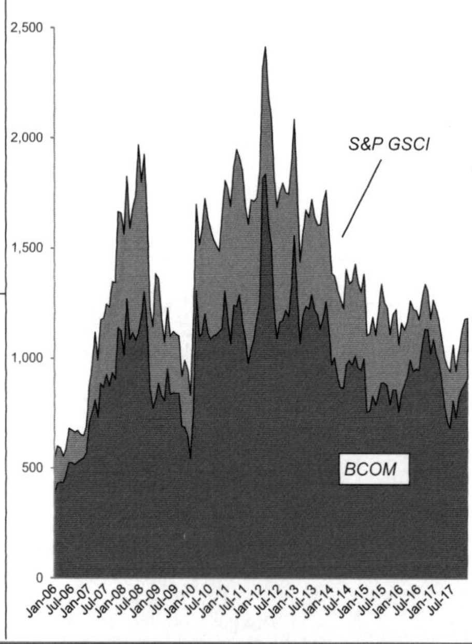

Source: January 2018 Zinc Market Outlook www.crugroup.com

CRU THE INDEPENDENT AUTHORITY
MINING | METALS | FERTILIZERS

We expect the market to return to balance in 2018 H2

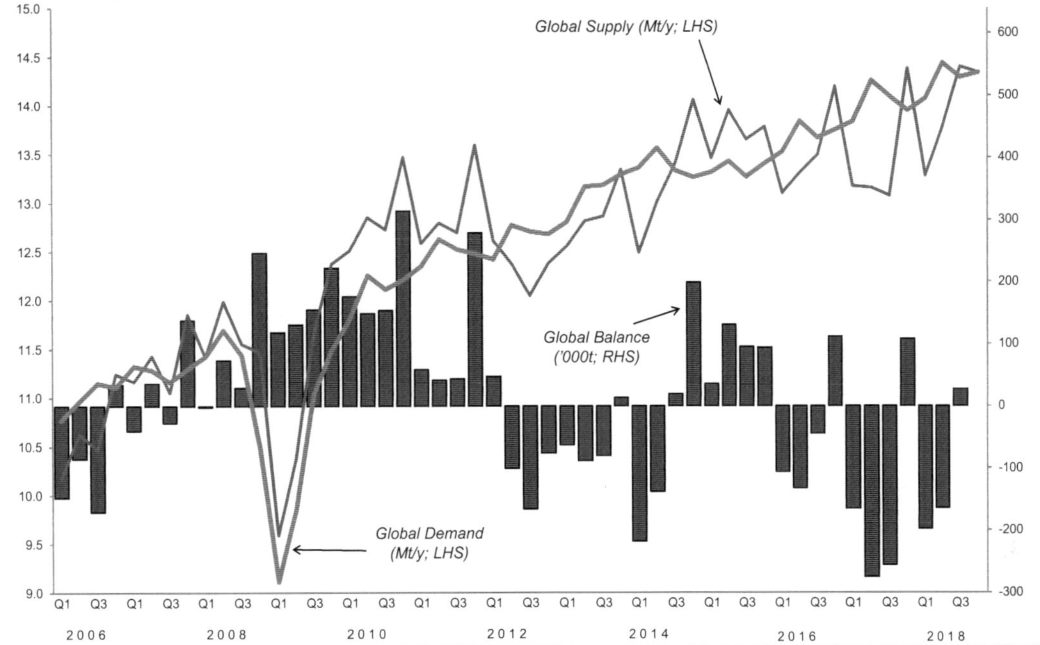

Source: January 2018 Zinc Market Outlook www.crugroup.com

CRU THE INDEPENDENT AUTHORITY
MINING | METALS | FERTILIZERS

Mine supply will continue to rebound

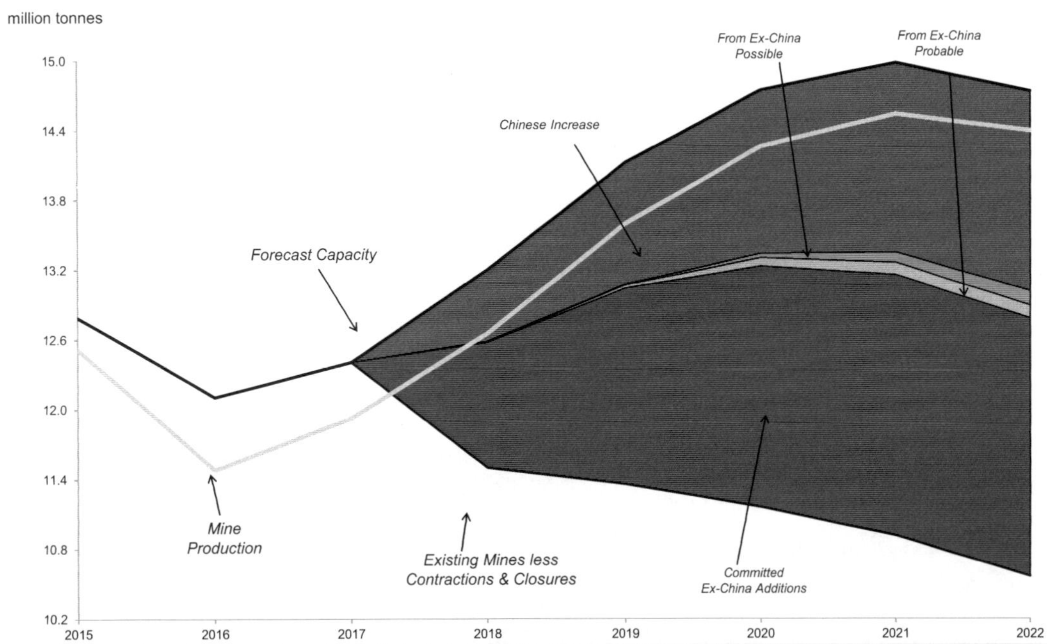

Source: January 2018 Zinc Market Outlook www.crugroup.com

CRU THE INDEPENDENT AUTHORITY
MINING | METALS | FERTILIZERS

Century Tailings will make a significant contribution

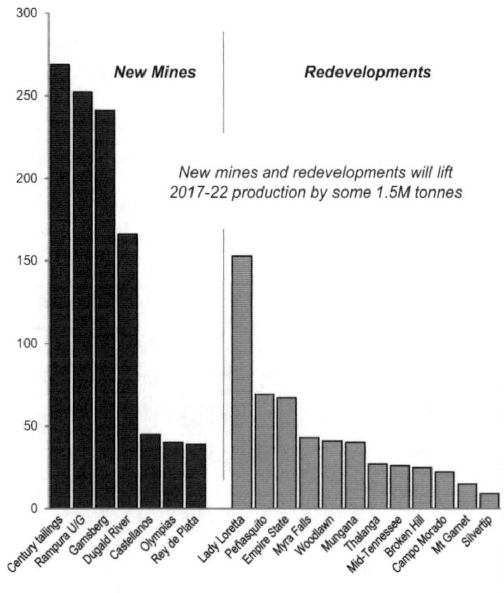

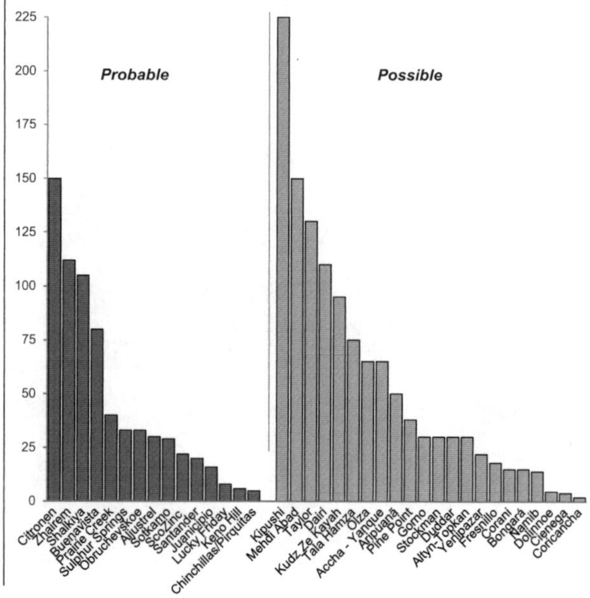

Source: January 2018 Zinc Market Outlook www.crugroup.com

CRU THE INDEPENDENT AUTHORITY
MINING | METALS | FERTILIZERS

Closures and contractions will cost over 1Mt

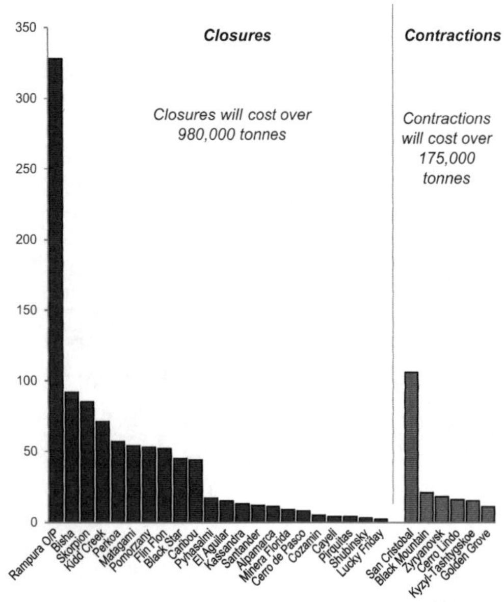

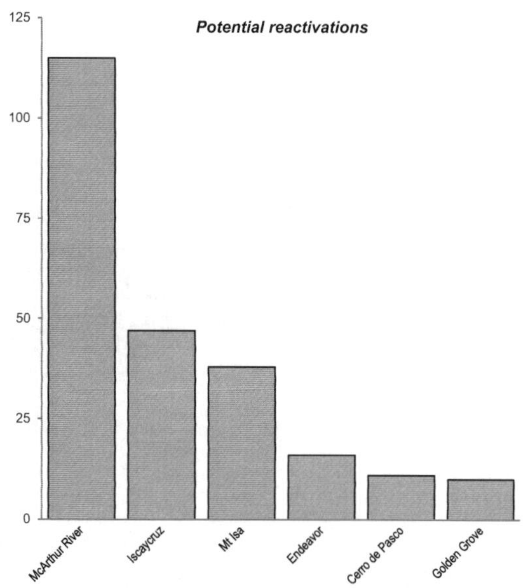

Source: January 2018 Zinc Market Outlook www.crugroup.com

Huoshaoyun could build to 200,000 t/y by 2022

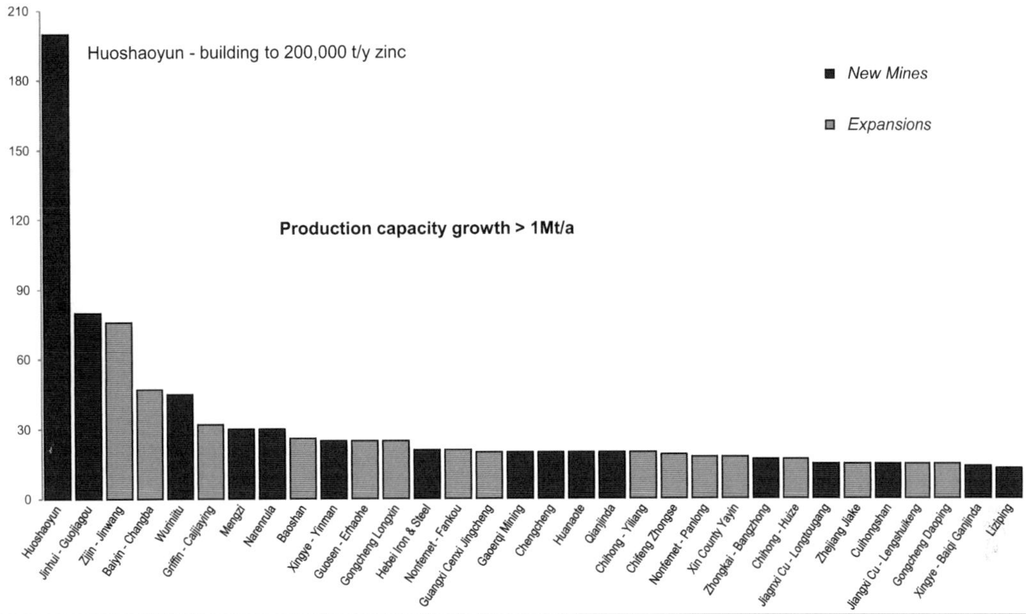

Source: January 2018 Zinc Market Outlook

Huoshaoyun is in a remote location

Source: CRU Insight December 2017

Guanghui Zinc intends to build smelters in Luopu

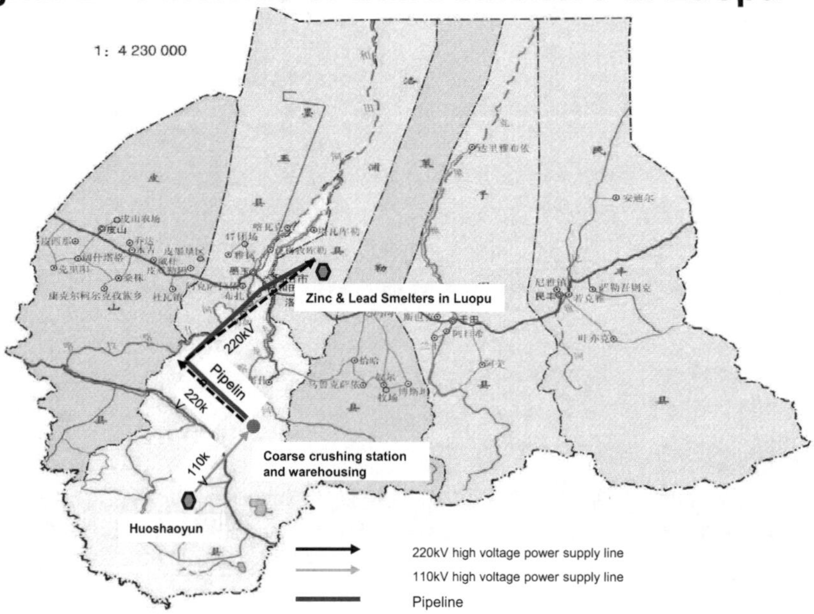

Source: Xinjiang Bureau of Surveying Mapping and Geoinformation

Debottlenecking is likely to be the greatest source of ex-China smelter capacity growth

2017-22 change in annual capacity; '000 tonnes

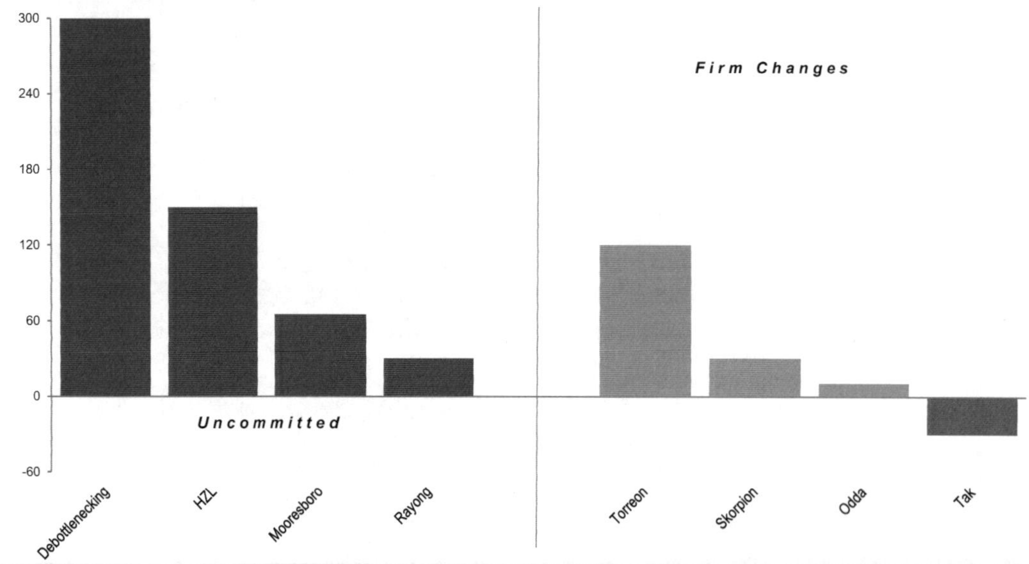

CRU THE INDEPENDENT AUTHORITY MINING | METALS | FERTILIZERS

Chinese smelter capacity growth is expected to average 3% in 2017-2022

2017-22 rise in nameplate capacity; '000 t/y

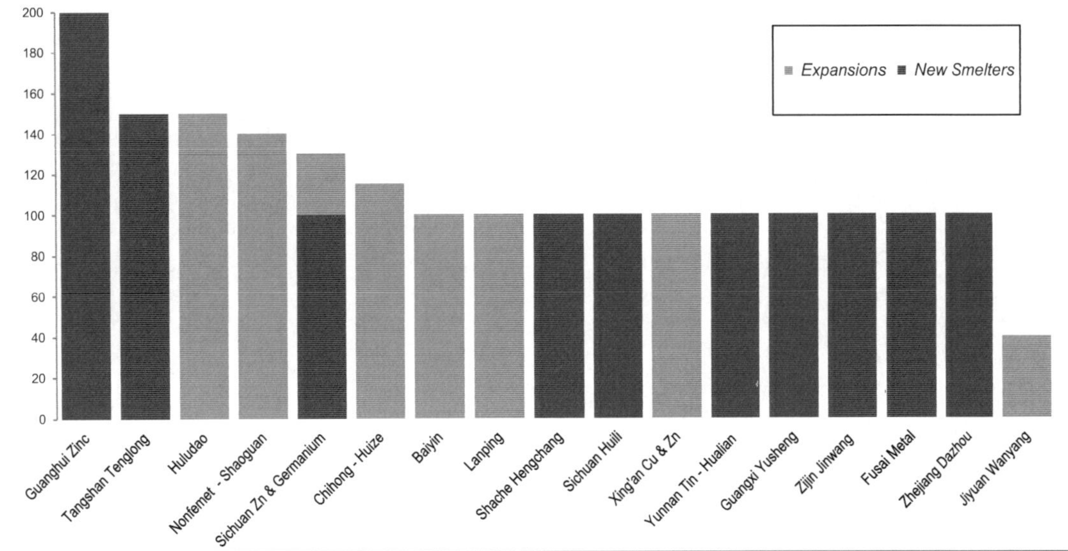

Source: January 2018 Zinc Market Outlook www.crugroup.com

CRU THE INDEPENDENT AUTHORITY MINING | METALS | FERTILIZERS

Global demand growth will slow in 2018 & 2019

refined zinc demand, million tonnes

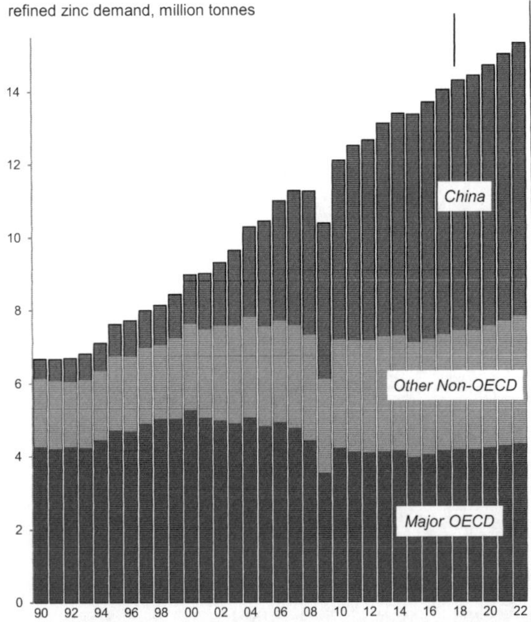

year-on-year percentage change in regional IP

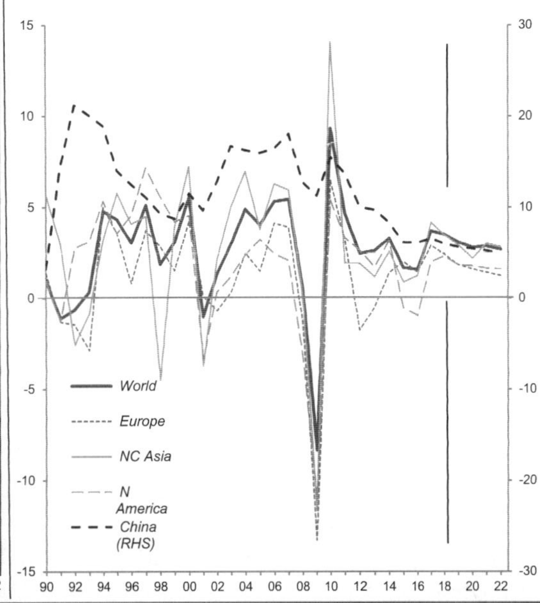

Source: January 2018 Zinc Market Outlook www.crugroup.com

Concentrate surpluses could become very large

cumulative zinc market surplus; adjusted; million tonnes

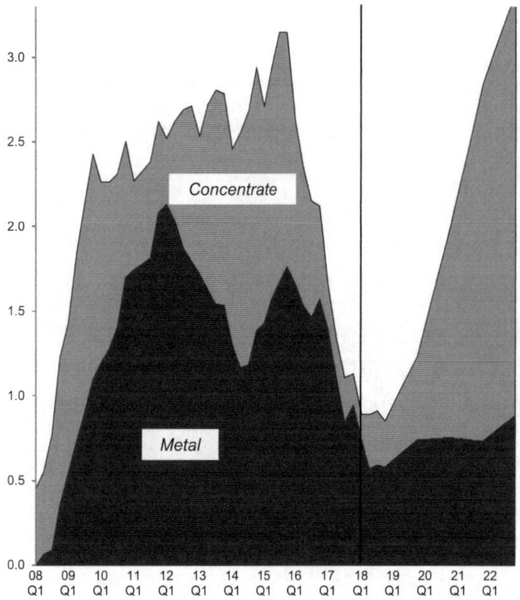

global market balances; '000 tonnes contained zinc

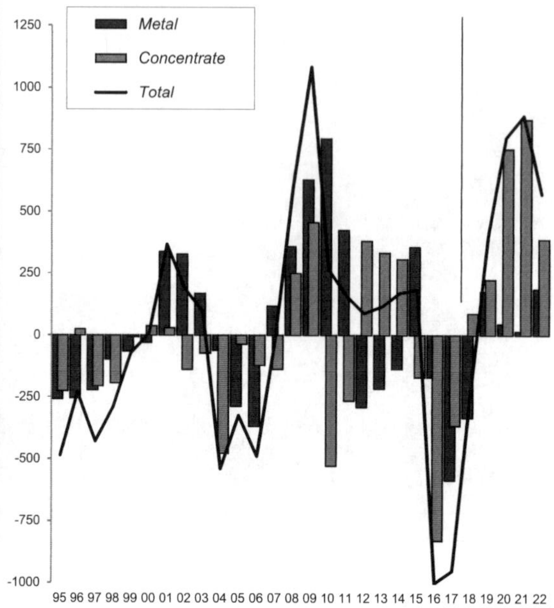

Source: January 2018 Zinc Market Outlook

www.crugroup.com

We expect prices to fall as stocks begin to build

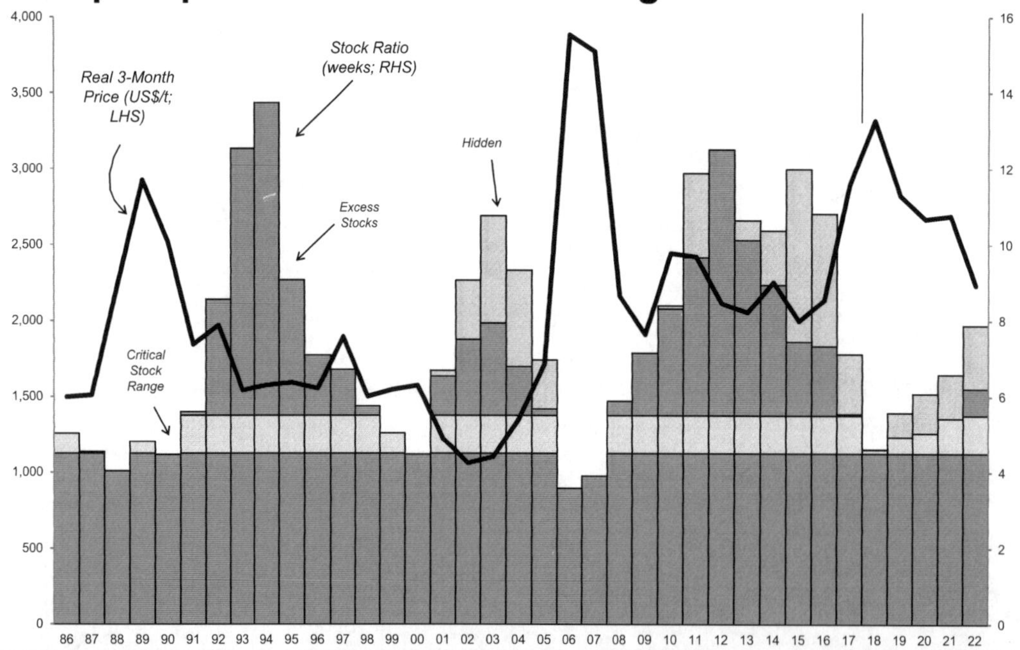

Source: January 2018 Zinc Market Outlook

www.crugroup.com

- Helen O'Cleary
  helen.ocleary@crugroup.com

- Christine Meilton
  christine.meilton@crugroup.com

- Ryan Cochrane
  Research Manager: Lead, Zinc and Precious Metals
  ryan.cochrane@crugroup.com

Zinc: the story of a long heralded market rally

IZA International Zinc Conference

Carlsbad – 13th February 2018

Concord Resources Limited

Duncan Hobbs
+44 20 3879 3966
duncan.hobbs@concordltd.com

Luis Woolcott
+51 978 305925
luis.woolcott@concordltd.com

CONCORD

Disclaimer

The entire content of this document is strictly confidential and subject to international copyright laws and other intellectual property rights with all rights reserved. No part of this document may be redistributed to any other person in whole or in part, stored in a retrieval system, modified, linked into, republished, uploaded or transmitted in any form or by any means electronic or mechanical or by photocopying, scanning, recording or otherwise, without the express written permission of Concord Resources Limited ("Concord") and by accepting this document, you agree to keep confidential the information contained herein and not use it for any purpose other than evaluating the information contained herein.

The information contained herein has been obtained from and opinions are based upon sources that are believed to be reliable but Concord does not make any representation with respect to any information contained herein and it should not be relied upon. The use of any information herein does not imply any verification of such information. All opinions, forecasts and estimates reflect Concord's judgement when this document was prepared and are subject to change without notice. In addition, any strategies, projections and plans described in this document are subject to change at any time at Concord's sole discretion. The information and projections contained herein are not a guarantee of future results. Any terms herein are intended for discussion purposes only and are subject to the final terms of final, written agreements.

This document and its contents and any discussion thereof do not constitute any offer to any person nor constitute any legal, regulatory, tax or accounting device and the authors do not recommend to any person any transaction discussed or referred to herein.

Any liability of whatsoever nature and howsoever arising of Concord relating to the contents of this document is hereby expressly disclaimed to the extent permitted by law.

Concord Resources Limited

February 2018

Concord Resources Limited
Who we are

- Founded in 2015 Concord is a global resources merchant focused on the supply chain, financing and risk management of non-ferrous metals and associated raw materials, led by an experienced and entrepreneurial management team
- Concord is a global company, operating from its headquarters in London and through its principal offices in New York and Hong Kong. Our experienced team work with resource producers and consumers on spot transactions out to multi-year, structured agreements
- Concord's capabilities include global sourcing, marketing and management of all required logistics and finance to optimise our clients' supply chains from long-term partnership marketing arrangements to single container deliveries
- Concord brings considerable cross-commodity, structuring and asset optimisation experience of its founding team to work for the benefit of its customers and partners
- Concord is an independent company, which combines its founding management's experience with the commodity and financial industry expertise of its shareholders, including:
  - MKS Switzerland SA
  - Pala Investments Limited
  - Principals from Ospraie Management
  - Pinnacle Asset Management
- For more information please visit our website

CONCORD 2

Zinc: the story of a long heralded market rally

- 2nd Edition
  - Chapter 1: Strong promises, false starts
  - Chapter 2: Mine closures and metal deficit
  - Chapter 3: World economy expands, and fast
- 1st Edition (2005-2006) review
- 2nd Edition: Epilogue
- A lead actor: Peru review
- Concord Resources Limited

CONCORD 3

Zinc: the story of a long heralded market rally

- 2nd Edition
  - Chapter 1: Strong promises, false starts
  - Chapter 2: Mine closures and metal deficit
  - Chapter 3: World economy expands, and fast
- 1st Edition (2005-2006) review
- 2nd Edition: Epilogue
- A lead actor: Peru review
- Concord Resources Limited

CONCORD 4

A new narrative emerged in zinc following the global financial crisis of a looming shortfall in mine output…

CRU Zinc Market Outlook

- Oct 2009
  "Large market surpluses will give way to deficits by 2012 … but without … closures, smelter capacity will exceed mine supply and metal demand"
- Oct 2012
  "Deficits back in view … [the] switch … will take time, but be severe"
- Oct 2014
  "Is 2015 zinc's year? Metal demand has now moved above mine supply"

Macquarie Commodities Research

Zinc

In surplus today but outlook for galvanising metal begins to glimmer

2 October 2013

Glencore + Add to myFT

Metal bulls pin their hopes on zinc as mines close

Metal predicted to jump to more than $3,500 a tonne in 2016-18

Martin Sandbu in London OCTOBER 22, 2013

COMMODITIES

Mine Closures Are Set to Give Zinc a Boost

Zinc Prices Rose to a Nine-Month High, and Supply Is Expected to Tighten

By Rhiannon Hoyle And Francesca Freeman

Dec. 20, 2013 7:30 p.m. ET

Source: CRU, FT, MBL, WSJ, Concord Resources Limited, February 2018

CONCORD 5

Yet ageing zinc mines lived on and world output exceeded expectations, while demand disappointed

CRU **Zinc** Market Outlook

- Oct 2015: *"Were we wrong? [No…] forecast of falling stocks and rising prices is intact".*

REUTERS

#COMMODITIES JUNE 12, 2015 / 6:22 PM

Zinc's infinitely stretchable deficit deadline: Andy Home

Andy Home 7 MIN READ

Brunswick, Century & Lisheen mines – cumulative output from 2011-2015

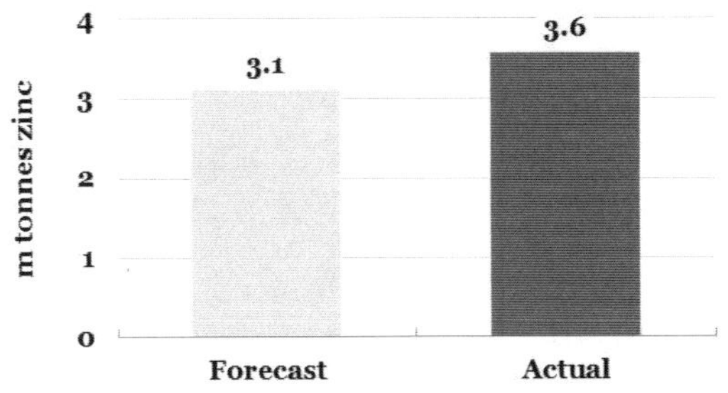

World zinc consumption growth from 2011-2015

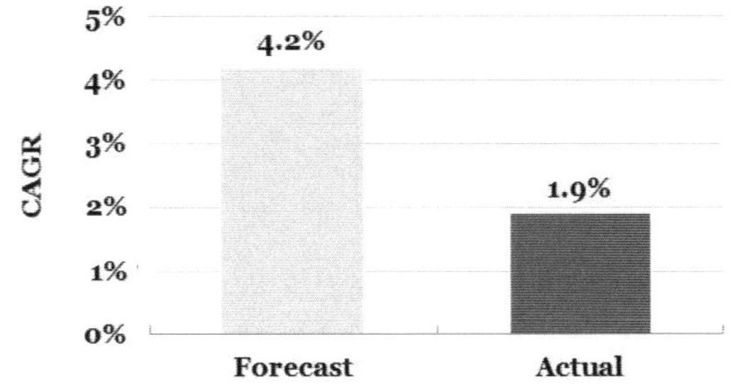

Note: Charts show average forecasts from selected sources for the period 2011-2015, as published in early 2011, and actual outturn for the period according to latest published data
Source: CHR, CRU, WM, Concord Resources Limited, February 2018

And zinc prices traded an established range, before breaking *down* in late 2015

LME zinc price

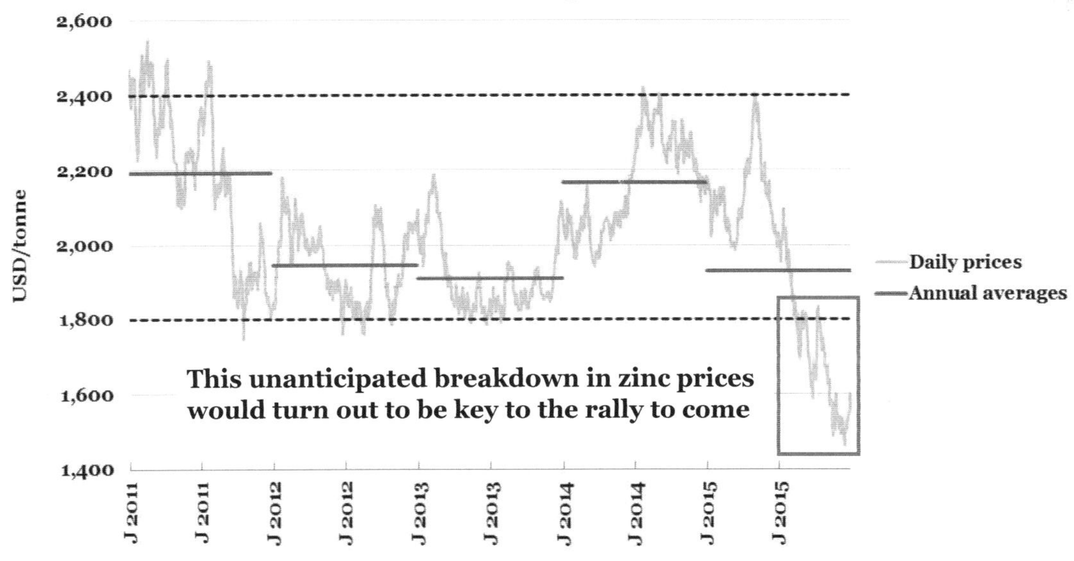

Note: cash settlement prices
Source: LME, Concord Resources Limited, February 2018

Zinc: the story of a long heralded market rally

- 2nd Edition
  - Chapter 1: Strong promises, false starts
  - Chapter 2: Mine closures and metal deficit
  - Chapter 3: World economy expands – fast
- 1st Edition (2005-2006) review
- 2nd Edition: Epilogue
- A lead actor: Peru review
- Concord Resources Limited

Finally key zinc mines closed due to resource depletion; at the same time some miners *voluntarily* cut production

MMG WE MINE FOR PROGRESS

Final shipment of Century zinc concentrate on its way

14 Dec 2015 12:00 AM Media Release

Lisheen mine in Tipperary makes final shipment

The mine, which began operations in 1999, employed up to 400 people at full production

Mon, Jan 25, 2016, 12:37

Charlie Taylor

theguardian

Glencore slashes zinc production and jobs

Friday 9 October 2015 09.01 BST

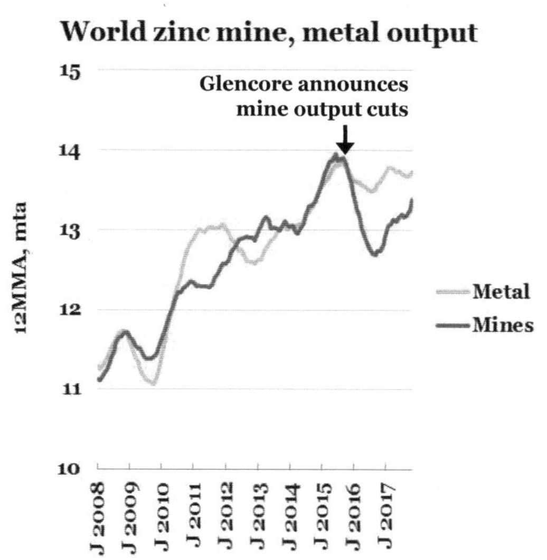

- Shortfall of mine output vs metal production implies de-stocking of zinc concentrates

Note: Chart data do not account for smelter yield losses nor secondary zinc feeds, but these net off to some extent
Source: IZSG, MMG, The Guardian, The Irish Times, Concord Resources Limited, February 2018

The zinc metal market tipped into deficit; reported stock levels have been drawn down sharply

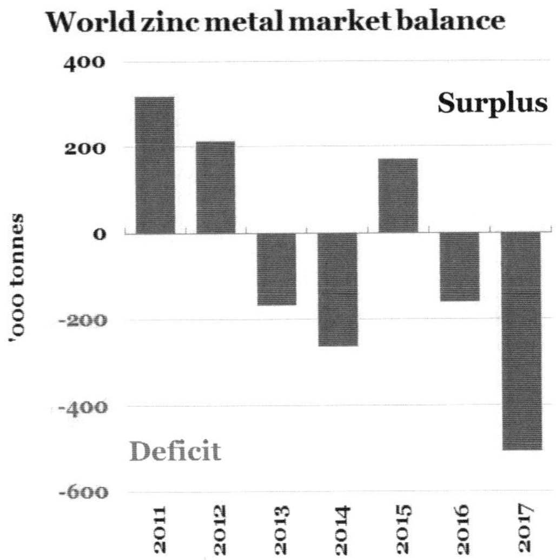

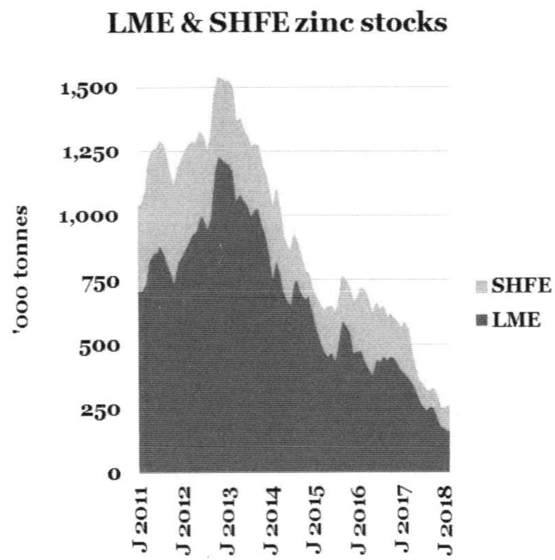

- But unreported metal stocks have buffered the market; debate now about what's left…

Note: 2017 market balance annualised basis published data Jan-Nov. Stock data as at month end, except latest
Source: ILZSG, LME, SHFE, Concord Resources Limited, February 2018

Zinc: the story of a long heralded market rally

- 2nd Edition
  - Chapter 1: Strong promises, false starts
  - Chapter 2: Mine closures and metal deficit
  - Chapter 3: World economy expands, and fast
- 1st Edition (2005-2006) review
- 2nd Edition: Epilogue
- A lead actor: Peru review
- Concord Resources Limited

Industrial production accelerates to fastest rate of increase for years in synchronous recovery; world zinc consumption reaches new records

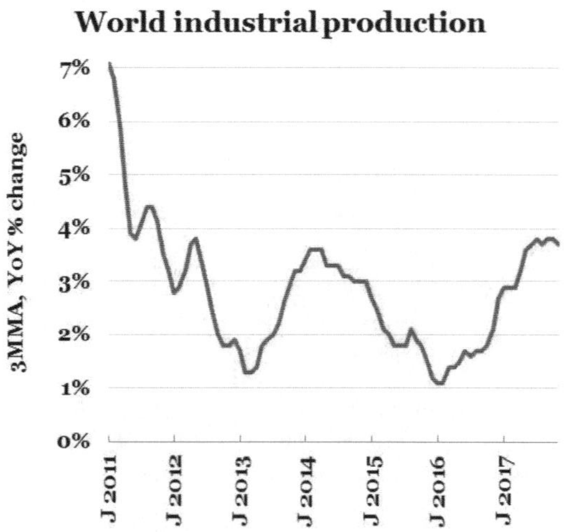

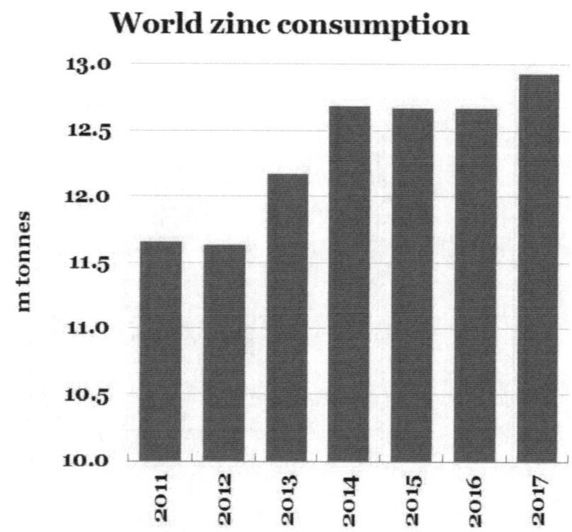

- China's IP may have been overstated during downturn; recovery may be stronger than it looks
- Zinc consumption recently rising faster than its long run rate (2017 +2.1% vs 10Y CAGR +1.8%)

Source: CHR, CPB, Concord Resources Limited, February 2018

Latest data point firmly to strong industrial production growth and buoyant metals demand continuing into 2018

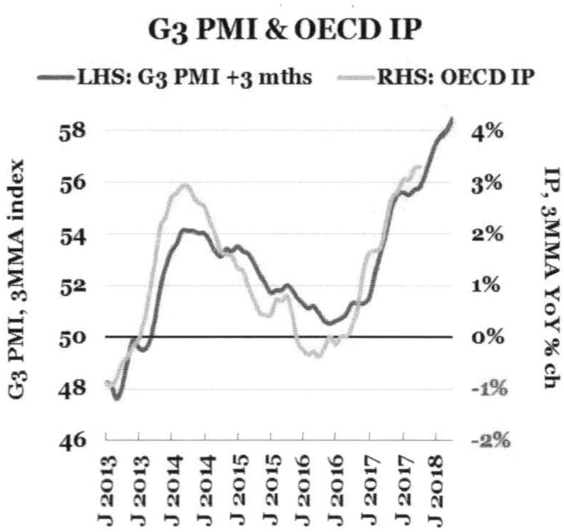

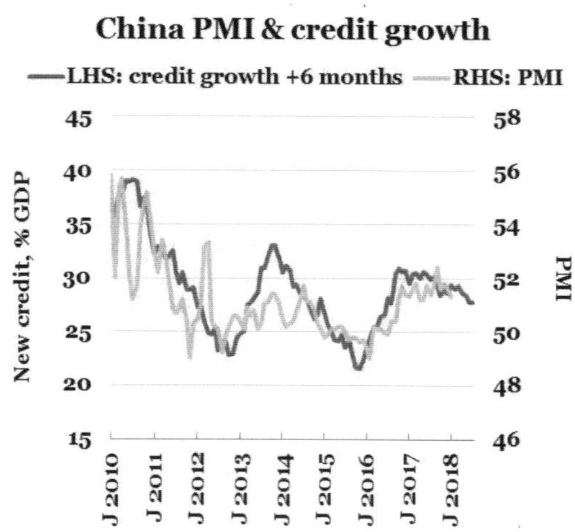

- 14 / 15 of world's largest industrial economies scored PMIs > 50 in Jan, but inflation building
- China's latest data point to stable outlook; no sign of any significant slowdown at this stage

Note: G3 PMI – USA, Eurozone & Japan weighted by industrial production. China's new credit calculated as 12-month flow of TSF, net of equity issues, plus local govt bond issues
Source: BBG, ISM, Markit, NBS, OECD, Concord Resources Limited, February 2018

The world has entered into an upgrade cycle, with positive implications for market sentiment

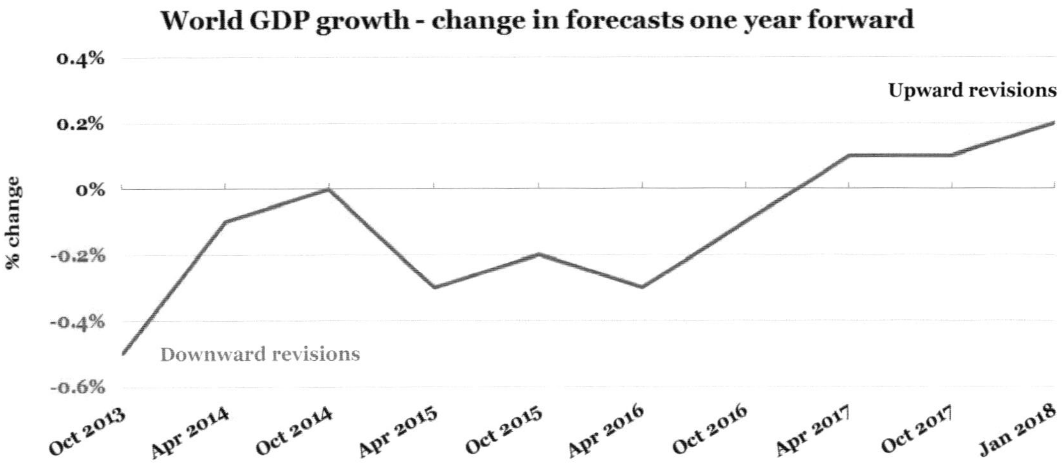

- Following years of disappointment, economic growth forecasts are now being raised
- Upgrade cycle animates 'animal spirits' – the recent jolt in equity markets aside!

Note: Chart shows change in forecast for year following date of issue vs that published in the immediately preceding period; for example, forecast for 2018 in Oct 2017 vs Apr 2017
Source: IMF, Concord Resources Limited, February 2018

Zinc: the story of a long heralded market rally

- 2nd Edition
  - Chapter 1: Strong promises, false starts
  - Chapter 2: Mine closures and metal deficit
  - Chapter 3: World economy expands – fast
- 1st Edition (2005-2006) review
- 2nd Edition: Epilogue
- A lead actor: Peru review
- Concord Resources Limited

Zinc price rallies outperformed the broader metals market in both the mid-2000s and from late 2015, but the latter to a lesser extent

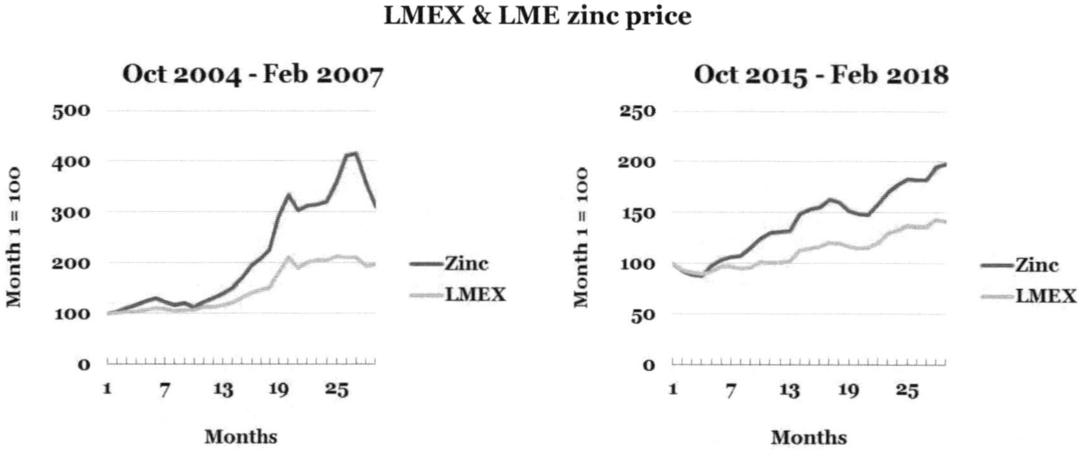

- LME zinc price more than doubled vs the broader market at its peak in the mid-2000s
- LME zinc has outperformed the broader market in the latest rally by ~50%

Note: All data are month averages basis official cash settlement prices
Source: LME, Concord Resources Limited, February 2018

CONCORD 16

Benchmark C-3M LME zinc spreads saw steeper backs in mid-2000s than have been seen so far in latest rally

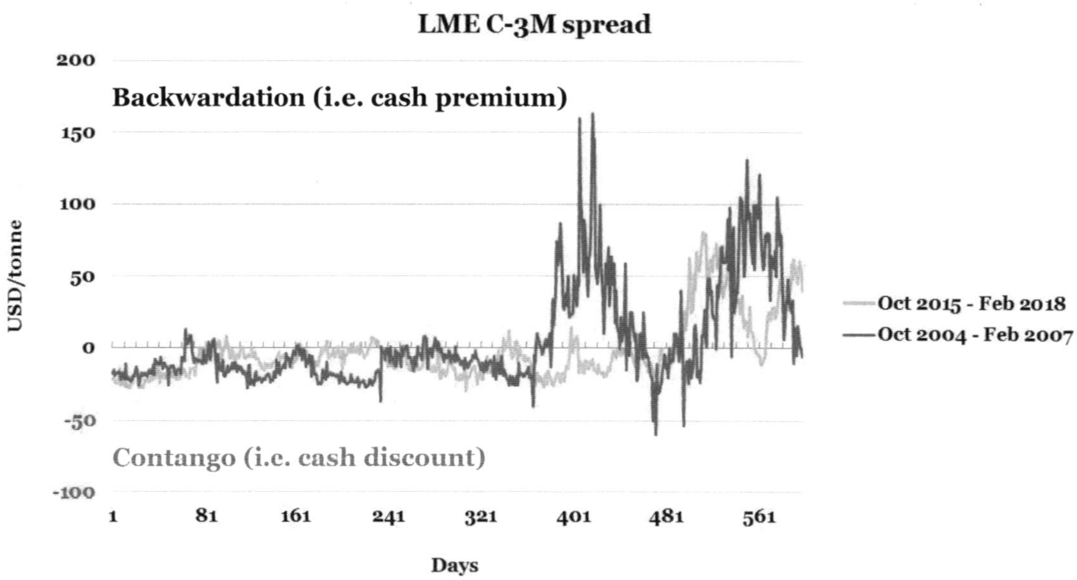

Note: All data basis official settlement prices
Source: LME, Concord Resources Limited, February 2018

CONCORD 17

Spot premiums for physical metal more than doubled in the mid-2000s; the last two years have seen only a small uptick at best

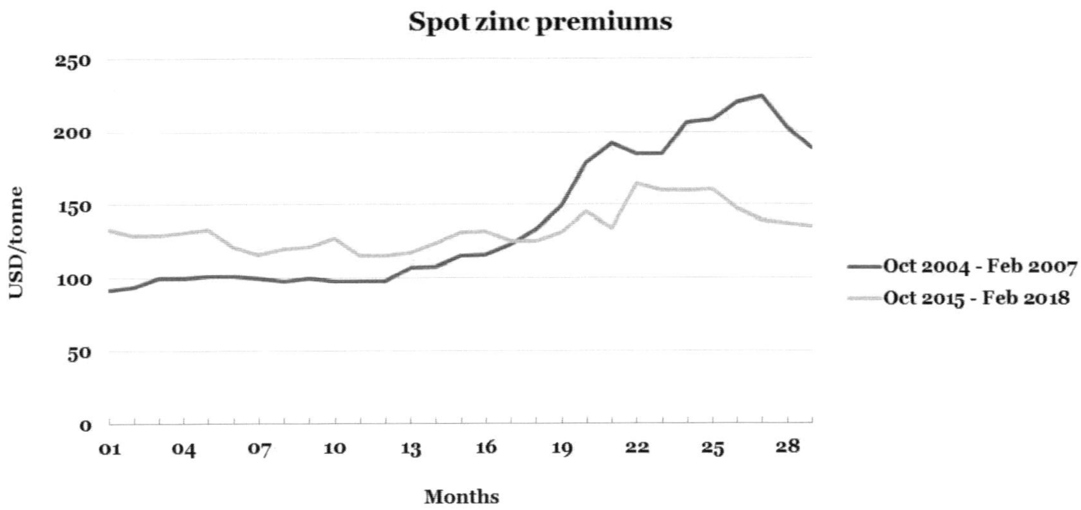

- In the latest cycle part of the uptick in premiums was due to a strike at a key smelter in Canada

Note: Chart shows consumption weighted average of spot premiums for SHG ingot CIF China; DDP US Midwest; EXW NW Europe
Source: CRU, LME, MB, Concord Resources Limited, February 2018

Zinc TCs have fallen to lower levels for a longer time since late 2015 than seen in the mid-2000s…

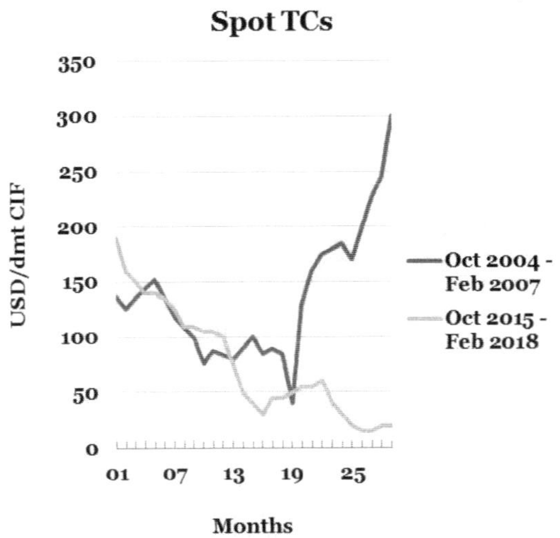

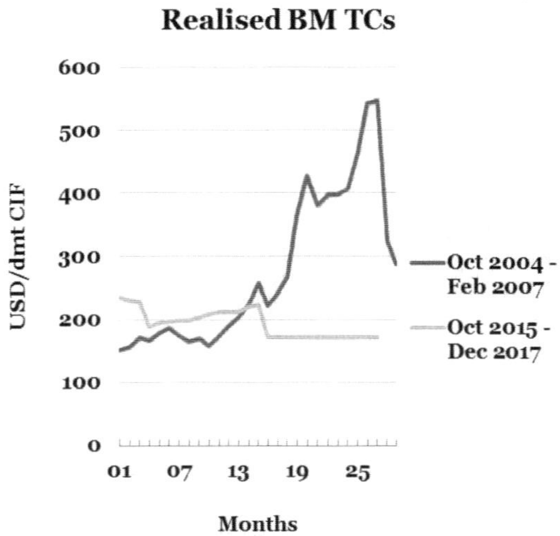

- BM TCs flat over the last year after up and down scales in price participation were set to zero

Note: TCs for typical clean concentrates ±53% zinc contained on standard payment terms. BM = "benchmark". 2018 BM terms not finalised at time of writing
Source: CRU, LME, MB, Concord Resources Limited, February 2018

And miners' share of the zinc price has risen to record levels over the last year in both the spot market and on BM

Miners' share of unit zinc price

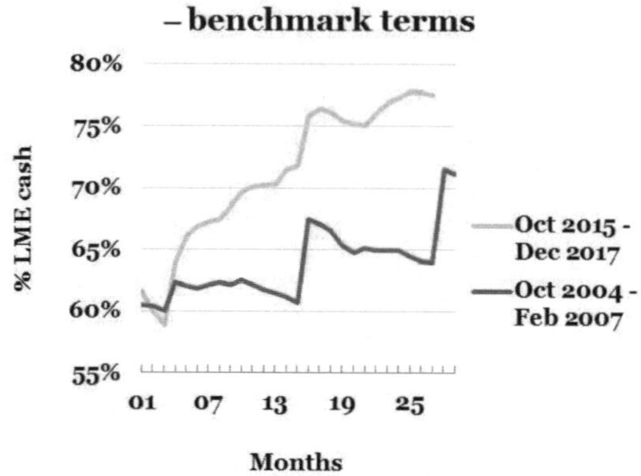

- Miners' price share on BM terms in the mid-2000s was kept lower by steep escalators in PP
- Adding metal premiums, which accrue to smelters, lowers miners' all-in price share by 3%-4%

Note: TCs for typical clean concentrates ±53% zinc contained on standard payment terms. 2018 BM terms not finalised at time of writing
Source: CRU, LME, MB, Concord Resources Limited, February 2018

We see very different trends in world zinc mine output in the mid-2000s vs the last two years

World zinc mine output

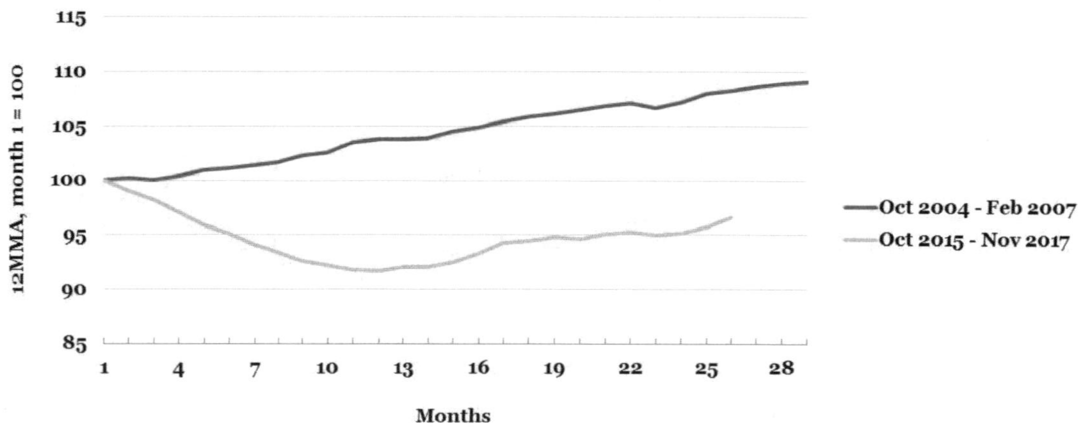

- 12MMA in mine output up ~10% in 2 years to early 2007; down >8% from late 2015-2016
- Note, however, world zinc demand was rising in the mid-2000s but soft in 2015-2016

Source: ILZSG, Concord Resources Limited, February 2018

Zinc: the story of a long heralded market rally

- 2nd Edition
  - Chapter 1: Strong promises, false starts
  - Chapter 2: Mine closures and metal deficit
  - Chapter 3: World economy expands – fast
- 1st Edition (2005-2006) review
- 2nd Edition: Epilogue
- A lead actor: Peru review
- Concord Resources Limited

CONCORD 22

Rebound in world mine output is widely expected to tip balances in world zinc market in 2018-2019

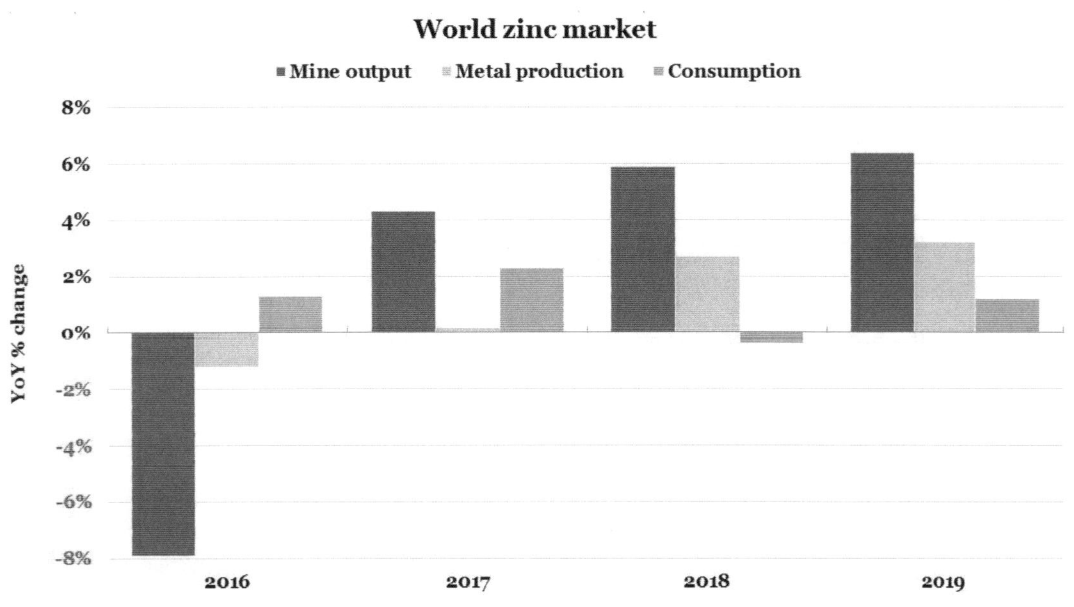

Note: Chart data are averages of selected sources. 2018-2019 data are forecasts
Source: CHR, CRU, WM, Concord Resources Limited, February 2018

CONCORD 23

In 2017, China's zinc miners added only a modest amount to output vs very strong price incentive

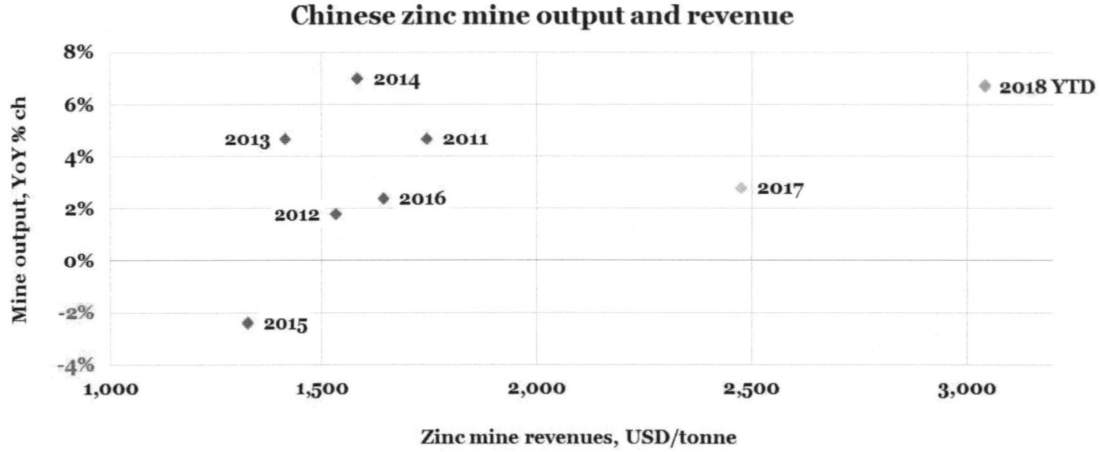

- 2018 is forecast to be very different with a large addition to local mine output
- The outcome will have a critical bearing on the world zinc market balance

Note: All data are annual averages, unless stated. Mine revenue per tonne of zinc metal production, calculated basis spot TCs on standard payment terms
Source: CHR, CRU, LME, Concord Resources Limited, February 2018

CONCORD

Inflationary worries – albeit absent much evidence as yet – are rising, central bank policies are changing and the outcome is uncertain

The QE2, former flagship of Cunard Lines, ran aground on a sandbank in The Solent approaching Southampton harbour on her final voyage from New York to the UK in November 2008, two months after the collapse of Lehman Brothers, an event emblematic of the Global Financial Crisis that left governments everywhere struggling to steer a safe economic course. Seen here are former UK prime minister, Gordon Brown, captain at the rudderless wheel, and former finance minister, Alistair Darling, raising the white flag as boom turned to bust...

Source: The BBC, The Daily Telegraph, Concord Resources Limited, February 2018

- A whole flotilla of QE policies followed the GFC to refloat the world economy, with a profound impact on asset allocation and prices
- With inflation worries rising, QE policies are being decommissioned, and interest rates are rising
- Maybe the world economy will not be beached by this change of course, but it may not be plain sailing into headwinds of high debt and increased inequality...
- And the impacts on asset allocation and prices – commodities such as zinc included – in these uncharted waters is highly uncertain

CONCORD

LME metals prices have tended to trade in tandem with inflation expectations and yields in recent years

- But correlation is *not* necessarily causation; past performance no guarantee of future outcomes

Source: BBG, LME, Concord Resources Limited, February 2018

CONCORD 26

Zinc: the story of a long heralded market rally 2nd Edition: Epilogue

- For some years following the GFC, zinc prices traded an established price range as output at key mines exceeded expectations and demand disappointed
- When mine closures due to resource depletion finally came the impact was amplified by voluntary cuts in response to an unanticipated *breakdown* in prices in 2015
- A deficit in the metal market has duly followed, owing in part to strong growth in the world economy driving demand, but this has been well buffered by stocks to date
- The 2nd Edition of the zinc price rally has some weaker plot lines than the first in the mid-2000s (premiums, spreads) but others are stronger (TCs, mine output)
- World zinc mine output is expected to accelerate well ahead of demand from H2 2018, tipping the balance in the zinc market, but China remains a wild card
- Broader macro risks may overwhelm zinc market 'fundamentals', perhaps driving the rally to new (even unanticipated) highs, or crushing expectations...

CONCORD 27

Zinc: the story of a long-heralded market rally

- 2nd Edition
  - Chapter 1: Strong promises, false starts
  - Chapter 2: Mine closures and metal deficit
  - Chapter 3: World economy expands – fast
- 1st Edition (2005-2006) review
- 2nd Edition: Epilogue
- A lead actor: Peru review
- Concord Resources Limited

CONCORD 28

Peru – present mining operations and projects

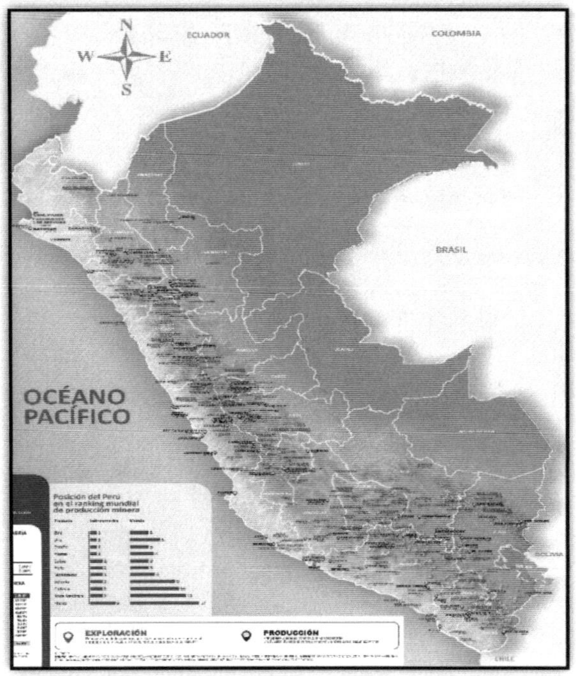

Source: Peruvian Mining Ministry, Concord Resources Limited, February 2018

Peruvian economy and mining industry

- Over 10% of Peru's GDP is generated by the mining industry
- 50% of the country's foreign exchange income comes from mining
- 20% of the country's tax take comes from the mining industry
- Peru's economy has transitioned from "high growth" (6%-9%) to "moderate growth" (2%-3%). 2017 estimated at ~2.5%. 2018 forecast at ~3.5%
- Key governmental investment programmes: Pope Francis' visit 2018 (January); "Niño Costero" reconstruction, Dakkar 2018; PanAmerican Games 2019
- Peru's political situation is still unsteady and probably will take some time to stabilise. This could affect both new mining and expansion projects in the country
- Logistics are constrained

Peru's evolution in sales & marketing of mine output: 1970s

- Mines owned *and financed* by government: Minero Peru Comercial
- Few trading companies
- Communications limitations
- Typically small / medium size mines
- Projects planned but tough to progress
- Smelters and refineries were also owned by government
- Logistics / Callao land owned by government
- In 1970s Peru supplied ~7% of world zinc mine output; by end of decade near 500ktpa zinc cont

centromin PERU

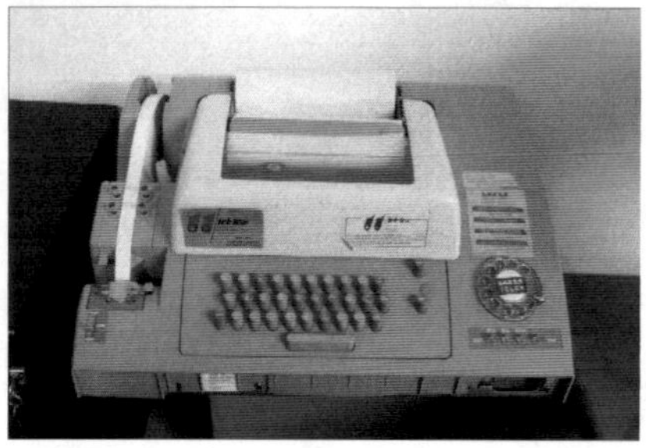

Source: ILZSG, Peruvian Mining Ministry, Concord Resources Limited, February 2018

Peru's evolution in sales & marketing of mine output: 1990s

- Mines and smelters / refineries privatised. End of Minero Peru Comercial
- New trading companies set up; some mines financed by long term agreements with trading companies
- Small, medium and large mines projects planned and started
- New communications systems
- Logistics improved and land around Callao leased to third parties
- In 1990s Peru supplied ~10% of world zinc mine output; by end of decade near 900ktpa zinc cont

Source: ILZSG, Peruvian Mining Ministry, Concord Resources Limited, February 2018

Peru's evolution in sales & marketing of mine output: 2010 to today

- Mines / smelters integrated, mining "houses" emerge, small / medium size mines integrate and expand
- Tenders
- Trading companies / asset investments
- Mines access bank financing, other sources
- Projects – more copper, less zinc
- Globalised communications
- Logistics developed but limitations persist
- In the current decade, output from Peru is supplying over 10% of the world total, now running at ~1.4mtpa zinc cont

Source: ILZSG, Peruvian Mining Ministry, Concord Resources Limited, February 2018

CONCORD 34

Where we are today and what to expect looking ahead

- Definitely, Peru is a mining country!
- Peru's miners and smelters make a key contribution to the world's base metals supplies
- Peru has changed the way it markets its mine products as the world has globalised
- Communications have played an important role
- Recently, projects have come to focus more on copper, and much less on zinc
- Peru's mine output should continue to increase; high metal prices are of course helping
- Logistics still constrained at some key points; more to be done here
- Don't forget to visit Peru!

CONCORD 35

The zinc market story

- **Thank you for your attention!**

- **Questions welcome to:**

  Duncan Hobbs
  +44 20 3879 3966
  duncan.hobbs@concordltd.com

  Luis Woolcott
  +51 978 305925
  luis.woolcott@concordltd.com

Zinc: the story of a long-heralded market rally

- 2nd Edition
  - Chapter 1: Strong promises, false starts
  - Chapter 2: Mine closures and metal deficit
  - Chapter 3: World economy expands – fast
- 1st Edition (2005-2006) review
- 2nd Edition: Epilogue
- A lead actor: Peru review
- Concord Resources Limited

Concord Resources Limited
Our markets: every step of the supply chain

Metals	Aluminum	Copper	Zinc, lead, nickel	Alumina, bauxite	Precious metals
				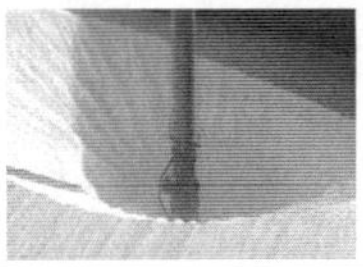	
Products	• P1020 • Billets • Alloys • Rod	• Cathode • Copper rod • Concentrates	• SHG & CGG zinc • 397 & 49s lead • Primary nickel • Concentrates	• SGA • Raw bauxite	• Gold • Silver • PGMs
Origination	• Consignments • Prepays/financing • Off-take/marketing • Tolling	• Consignments • Prepays/financing • Off-take/marketing • Tolling	• Consignments • Prepays/financing • Off-take/marketing • Tolling	• Swaps • Prepays/financing • Off-take/marketing • Tolling	• Off-take/marketing • By-products from concentrates
Storage / logistics	• LME system • Warehousing • Shipping	• LME system • Warehousing • Shipping • Blending	• LME system • Warehousing • Shipping • Blending	• Warehousing • Shipping • Blending	• Vaulting
Consumers	• Aerospace • Automotive • Construction • Packaging	• Wire and cable • Tubing • Brass • Smelters	• Galvanizers • Battery makers • Stainless & alloy steelmakers	• Refineries • Smelters	• Banks • Smelters

Concord Resources Limited
Contact our marketing teams

New York	London	Hong Kong
437 MADISON AVENUE 28TH FLOOR NEW YORK, NY 10022 USA	3 ST JAMES'S SQUARE LEVEL 8 LONDON, SW1Y 4JU ENGLAND	41 CONNAUGHT ROAD CENTRAL LEVEL 16 NEXXUS BUILDING HONG KONG
Scott Evans Senior Trader Americas Tel: +1 212 822 9793 scott.evans@concordltd.com	David Freeland Senior Trader EMEA Tel: +44 20 3879 3995 david.freeland@concordltd.com	Felix Cauro Senior Trader Asia Tel: +852 3757 9915 felix.cauro@concordltd.com

IHS Markit

Zinc Oxide Market Overview

Sean Davis, Principal Analyst, +1 408 343 4808, Sean.Davis@ihsmarkit.com

February 14, 2018

Operating rates improving?

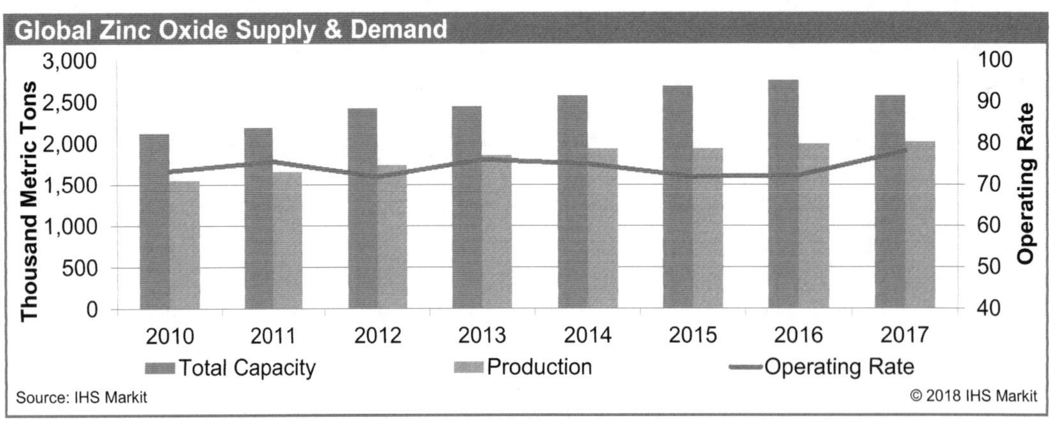

IHS Markit 3

Global Demand Summary

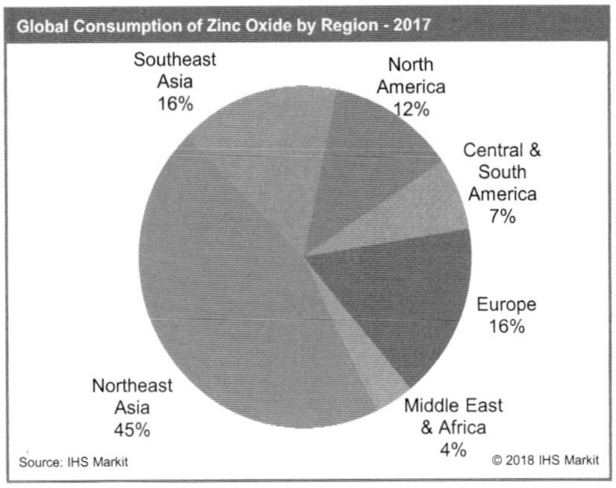

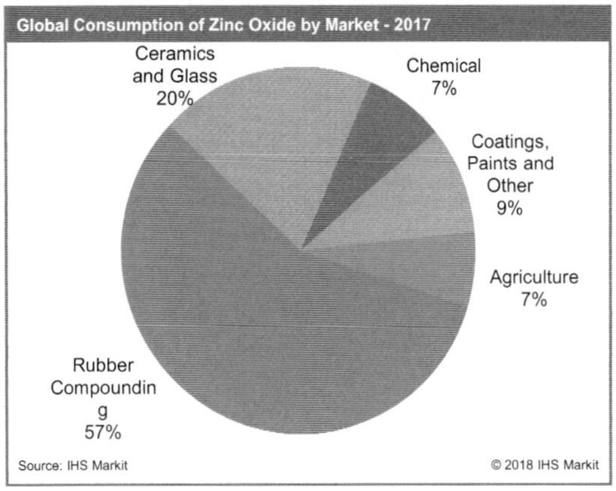

1.86 million metric tons

IHS Markit

Real GDP growth in major economies

Real GDP Percent change	2015	2016	2017	2018	2019
World	2.9	2.5	3.2	3.2	3.1
United States	2.9	1.5	2.2	2.5	2.2
Canada	0.9	1.5	3.0	2.3	2.2
Eurozone	2.0	1.8	2.4	2.1	1.8
United Kingdom	2.3	1.8	1.5	1.1	1.2
China	6.9	6.7	6.8	6.5	6.2
Japan	1.1	1.0	1.6	1.1	0.7
India*	8.0	7.1	6.8	7.4	7.6
Brazil	-3.8	-3.6	0.9	2.1	3.3
Russia	-2.8	-0.5	1.8	2.1	1.8

* Fiscal years starting 1 April
Source: IHS Markit

© 2018 IHS Markit

4

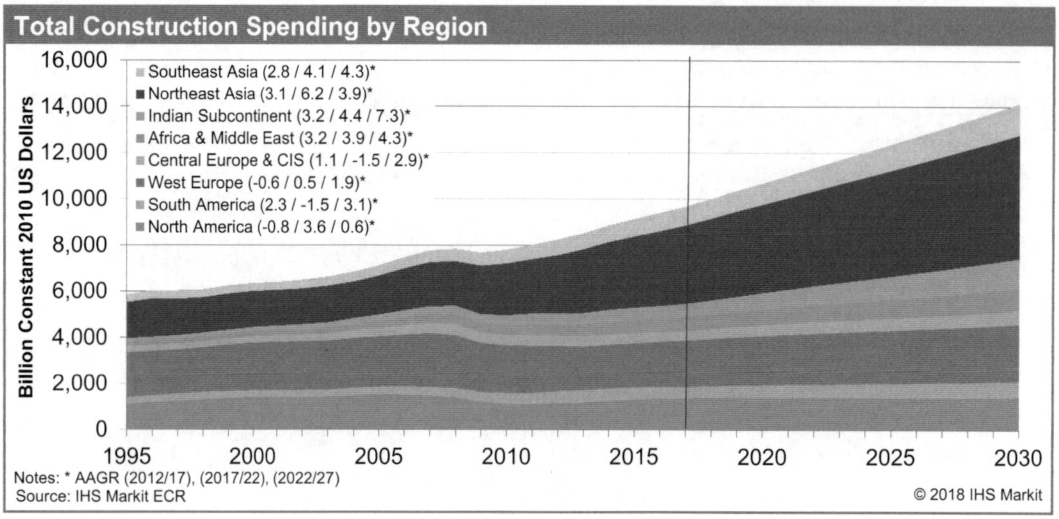

Zinc oxide applications can vary widely

- >Building / Industrial
- >Electronics
- >Energy
- >Mining
- >Consumer Goods

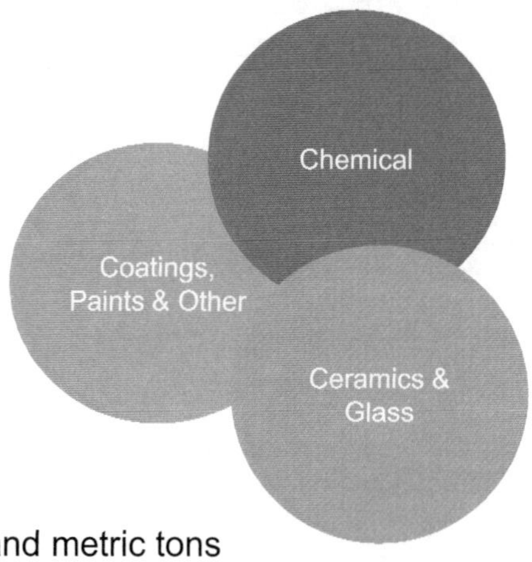

668 thousand metric tons

Zinc oxide's largest market segment

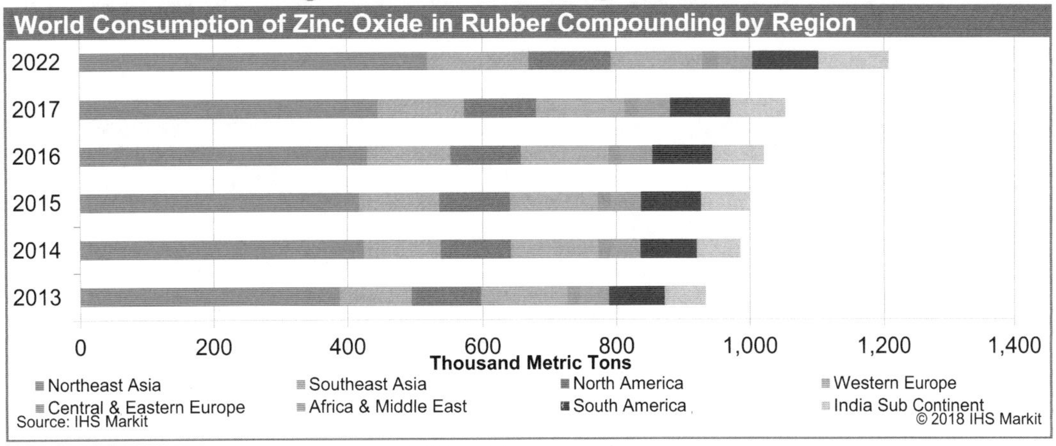

1.06 million metric tons

Potential demand from expanding tire capacity

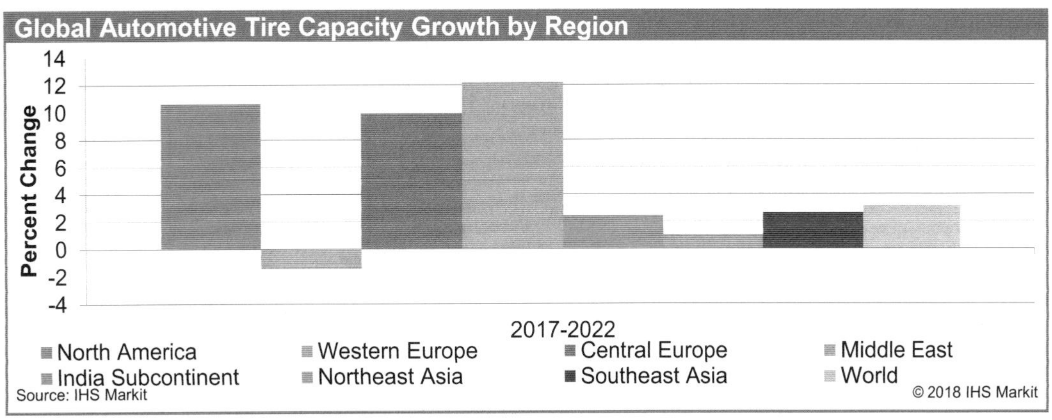

Slowing automotive production growth

Total Vehicle Production by Region

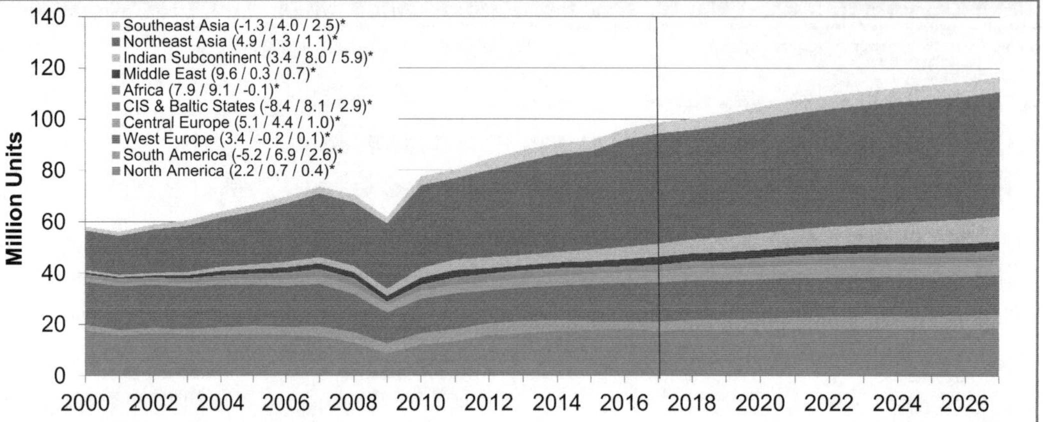

Notes: * AAGR (2012/17), (2017/22), (2022/27)
Source: IHS Markit

Major source in agriculture

Global Consumption of Zinc Chemicals in Agriculture - 2017

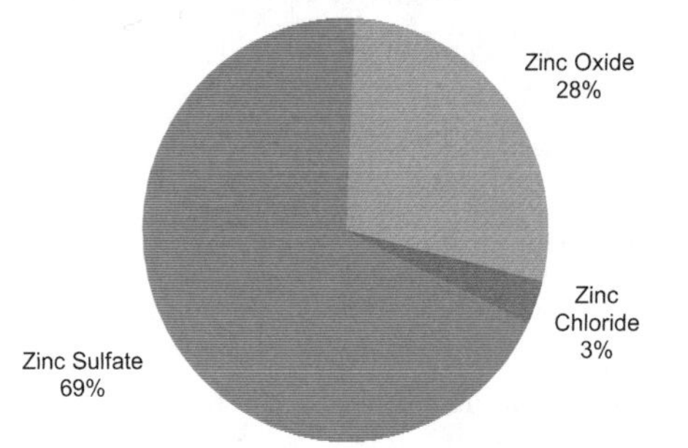

Source: IHS Markit

Consumption of Zinc Chemicals in Agriculture by Region - 2017

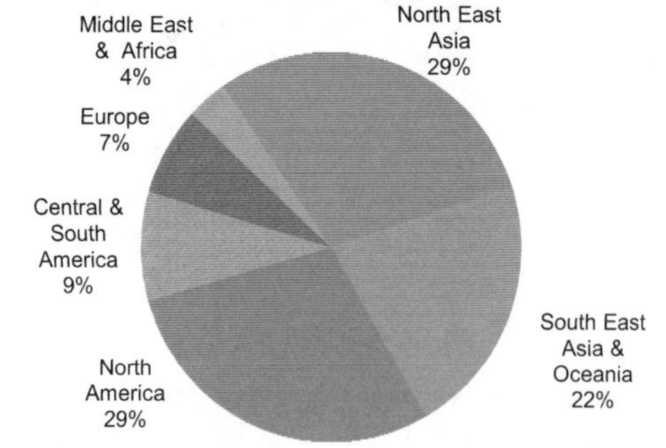

Source: IHS Markit

Zinc Oxide has a place in agriculture

$ZnSO_4$

Zn Oxysulfate

Zn Chelates

ZnO

Zn Sucrate

$ZnCl_2$

Zn Carbonate

Zn Lignosulfonate

Zn acetate dihydrate

Zinc Oxide in Agriculture is Improving

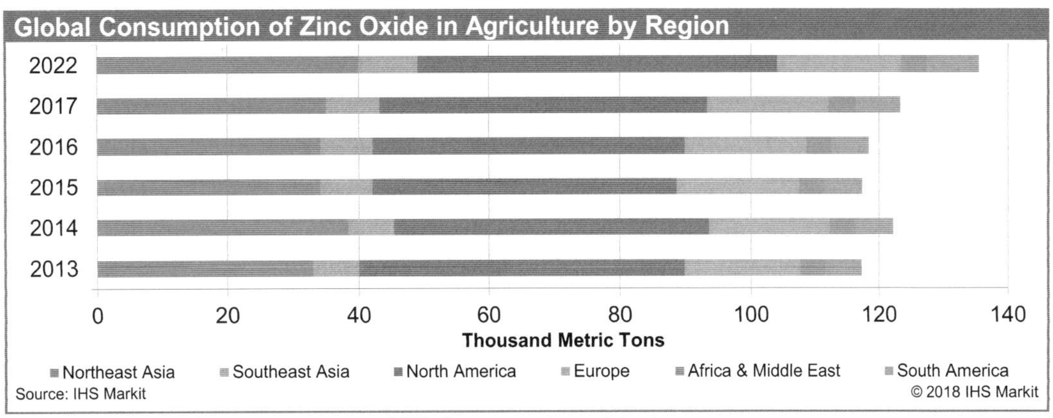

123 thousand metric tons

IHS Markit 13

Zinc oxide Agriculture Demand Growth to Steady

>The message is accepted - **Zinc works!**

>Technology will play an increasing role in managing inputs leading to a more tried and true approach to farming and nutrition management

>Record metal prices hinder growth

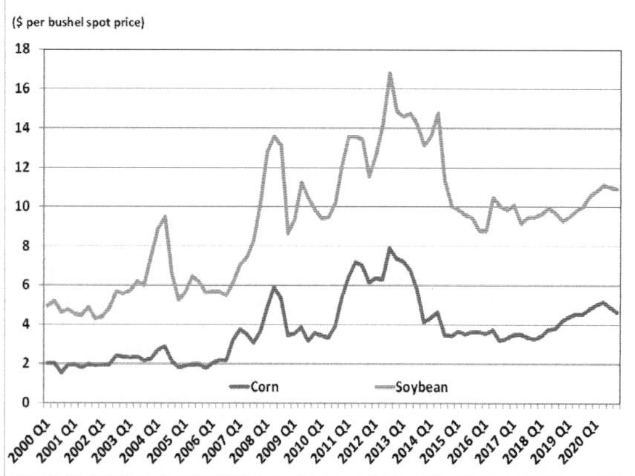

IHS Markit 14

Closing Remarks

>Favorable global economic growth fares well for zinc oxide

>Infrastructure spending is coming

>Changes in policy on trade pose risks

Is there a Reinvention of the Wheel

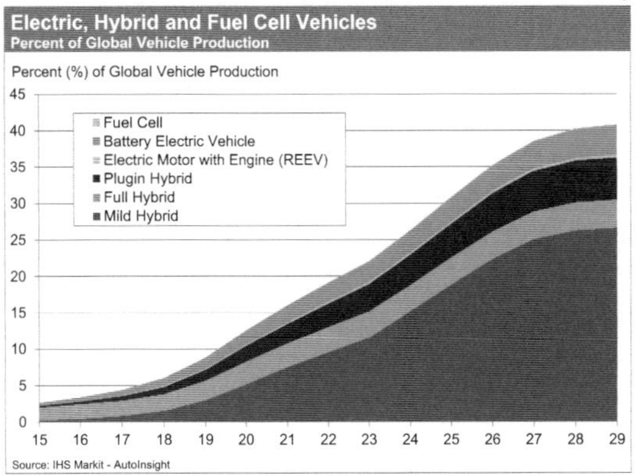

- Factors contributing to change:
  - Slowing automobile production growth and global oil consumption
  - Increasing populations concentrated in major urban centers
  - Shift in consumer purchasing patterns including auto ownership, increased ride sharing, etc.
  - Change in automotive drive train from combustion to hybrids to full electric
  - Autonomous vehicles
  - Legislation

Thank you!

IHS Markit Customer Care

CustomerCare@ihsmarkit.com
Americas: +1 800 IHS CARE (+1 800 447 2273)
Europe, Middle East, and Africa: +44 (0) 1344 328 300
Asia and the Pacific Rim: +604 291 3600

IHS Markit

Updates on Zinc Nutrient Initiative Activities

Dr. Andrew Green
Director EHS & ZNI, IZA

International Zinc Oxide Industry Conference

Carlsbad, USA
February 14, 2018

Bloomberg Markets Tech Pursuits Politics Opinion Businessweek Sign In Subscribe

Nutrient-Poor Farms Get a Vitamin Boost From Zinc Mines

By **Danielle Bochove**
April 17, 2017, 7:00 PM EDT *Updated on* April 18, 2017, 5:50 AM EDT

- **IHS Markit – 270,000 tpy market in 2017**
- **Potential is for 900,000 tpy**

International Zinc Association

ZNI Activities

CHINA: Key Developments

- **Organized over 200 demonstrations and training workshops in 20 provinces**
- **Over 20 fertilizer training workshops in fertilizer industry on Zn fertilizer technology**
- **Zn fertilizer education included in China National Agriculture Distant Education School Program reaching millions of agriculture extension workers and farmers**

International Zinc Association

Market Driving Food & Nutrition Security

- China program has increased Zn fertilizer use by 65,000 tons nationwide in 18 provinces
- This has resulted in:
  - Production of 40 million tons of Zn-enriched food
  - Improved nutrition for more than 200 million people
- Ministry of Agriculture has established China Zinc Rich Food Production Technology Committee

CHINA: Regulation and Policy Update

- Environment Protection organizations at all levels intensified inspections on fertilizer companies
- China developed "Environment Protection Tax"
- Impact: Many companies closed/reduced production; total fertilizer production reduced 9% from previous years
- China zinc sulphate production capacity declined to 600,000 tons in 2017

Success Story in India

- Addl subsidy in Zinc fortification included in NBSS
- **>200% increase in Zinc fertilizer use in India in past 6 years**
- 40 new grades of Customized Fertilizers containing Zinc
- Inclusion of ZnO susp, DAP-Zn, NPK-Zn, SSP-Zn, Zn-Poly Phos in FCO
- 45,000 tonnes $ZnSO_4$ mono plant by IFFCO + 16,500 tonnes $ZnSO_4$ hepta, mono plant by ANYA (KRIBHCO)

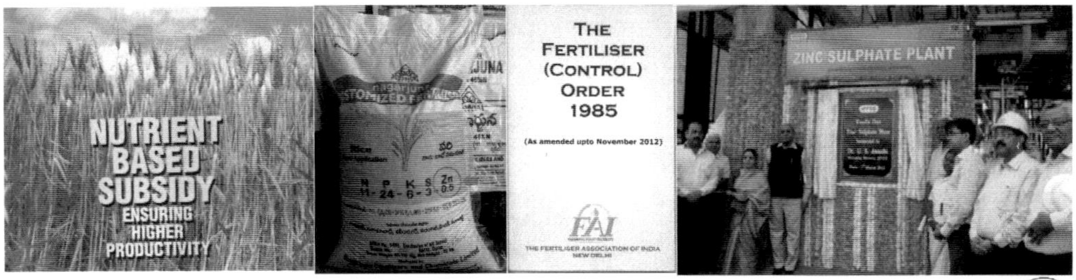

International Zinc Association

India Highlights on ZnO in Fertilizers

- New category of Zn-fertilizer is coated / fortified with ZnO, a preferred raw material (like in case of Zn-DAP, Zn-NPK, Zn-SSP)
- Zn-ash availability is decreasing = ZnO's demand increasing
- Zincated Urea when approved, will result in major boost for ZnO
- 2018 plan to include ZnO powder as fertilizer in FCO
- Support to develop new and innovative micronized / nano ZnO fertilizers

International Zinc Association

India Zinc Fertilizer Market

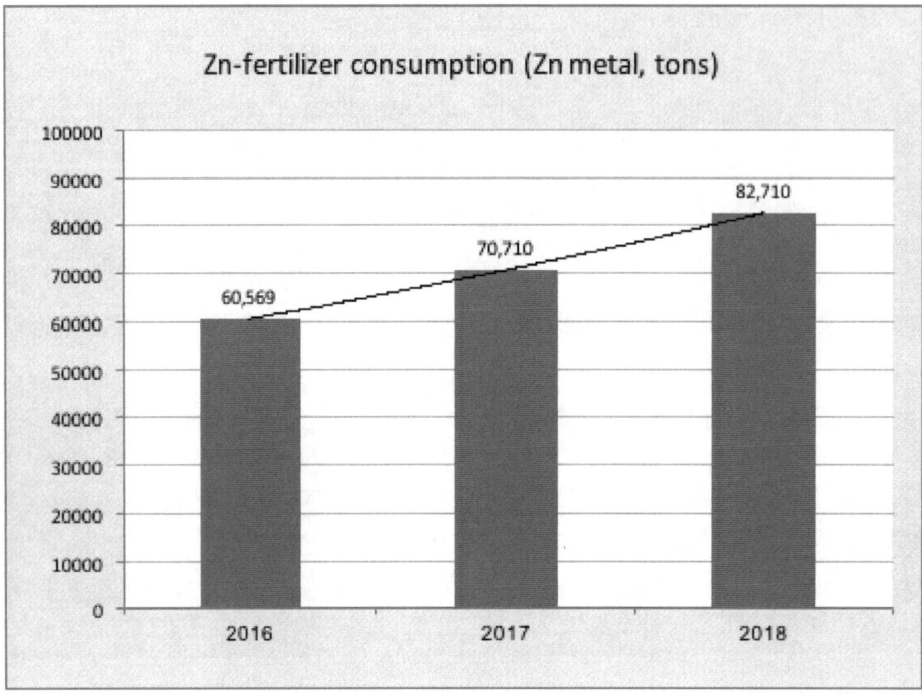

International Zinc Association

Key activities in 2018

- ZnO as Fertilizer in FCO
- Zincated Urea Policy (with ZnO coating / fortification)
- New innovative products (Micronized/Nano ZnO Fertilizers)
- GST on Zinc Fertilizers
- Zn Critical limit in soils
- Zinc in Soil-Plant-Animal-Human Continuum Research

International Zinc Association

Brazil: Activities

- Zinc deficiency map for Brazil
- Sugar cane crop trials

IAC
INSTITUTO AGRONÔMICO

- Fertilizer Seal of Approval

International Zinc Association

Zinc Fertilizer Seal of Approval

ZINC FERTILIZER SEAL OF APPROVAL

QUALITY VERIFIED

International Zinc Association

International Zinc Association

Zinc Fertilizer Seal of Approval

Table 1. Limits for impurities in Zn fertilizers

Element	Maximum allowed concentration per 1% Zn (mg/kg)	Maximum allowed concentration per kg Zn (mg element/kg Zn)	Maximum allowed concentration in ZnO (mg element/kg pure ZnO)	Maximum allowed concentration in $ZnSO_4$ (heptahydrate) (mg element/kg pure $ZnSO_4.7H_2O$)
As	60	6000	4820	1365
Cd	6.8	680	546	155
Cr	81	8100	6508	1843
Hg	130	130	104	30
Pb	5.3	530	426	121

International Zinc Association

Mexico: Activities

- Project with Instituto Nacional de Investigaciones Forestales,Agrícolas y Pecuarias (INIFAP)
- Workshops being held
- Discussions with government on policies
- **Market responding**
- **Zinc Saves Kids Initiative launched**

unicef PEÑOLES ZINC NACIONAL

International Zinc Association

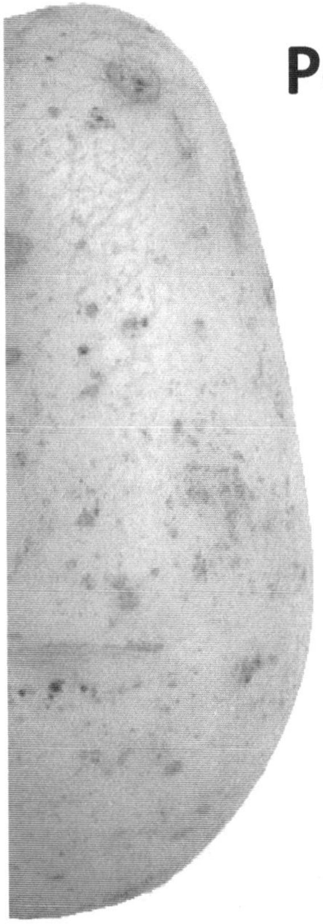

Peru Potato Project

- Working with the International Potato Institute and Universidad Nacional Agraria La Molina
- 4 trials underway

INTERNATIONAL POTATO CENTER
A CGIAR RESEARCH CENTER

- Goal is to improve potato production and nutrition

International Zinc Association

Bangladesh Project

- **50% of soils, 55% of people zinc deficient**
- **IZA-Bangladesh Agricultural Research Institute (BARI) Project on "Promotion of Zinc Fertilizer Use in Bangladesh for Food and Nutrition Security" underway**
- **11 crop trials, workshops and communications in progress**

Democratic Republic of the Congo

- 57% of population at risk for zinc deficiency
- Two programs implemented, funded by Lundin Foundation and MMG
- Focused on improving food & nutrition security with local farmers
- 23- 75% increase in grain production
- Value Cost Ratio of 11-31

5th International Symposium

ZINC

Improving Crop Production
And Human Health

Where: The Irish College, University of Leuven, Leuven, Belgium

When: Wednesday 5 September 2018 – Friday 7 September 2018

Organizers

international zinc association | ifa International Fertilizer Association | HarvestPlus | KU LEUVEN | Department of Earth and Environmental Sciences

International Zinc Association

http://www.zinccrops2018.org/

THE WORLD BANK
IBRD • IDA

Opportunity to add zinc to existing or planned World Bank agricultural activities

International Zinc Association

Crop Sector Projects by World Bank

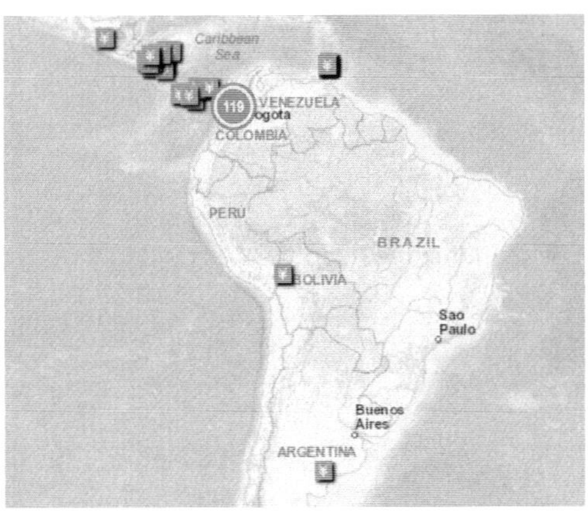

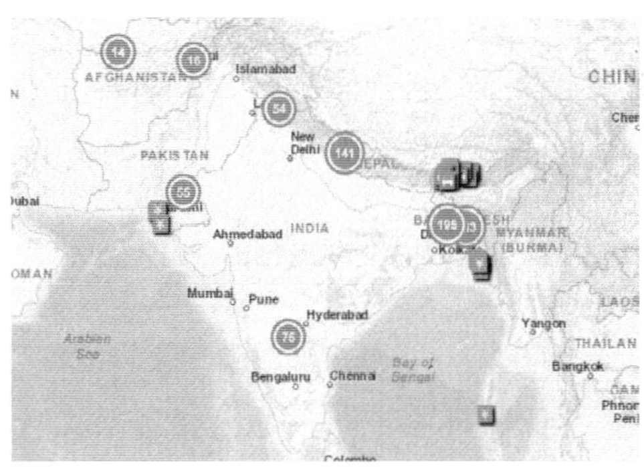

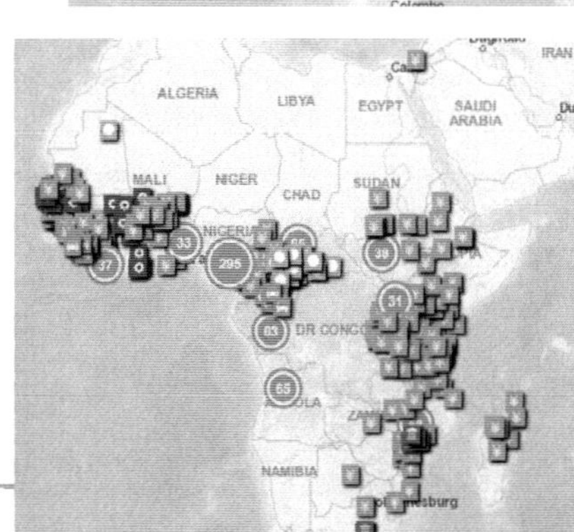

International Zinc Association

Zinc Fertilizer: Use and Potential

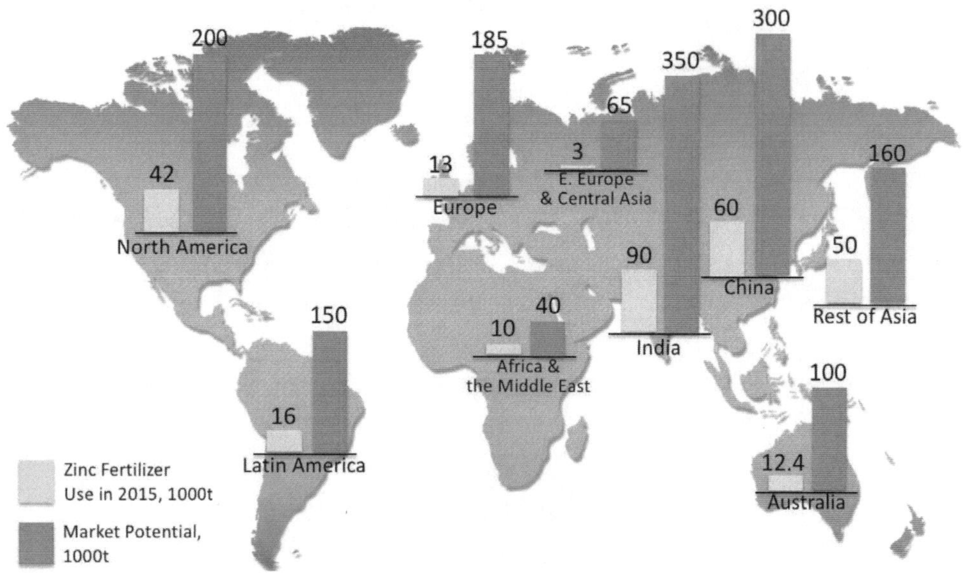

International Zinc Association

ZNI Affiliate Members

AMP
Advanced Micronutrient Products
CAMERON MICRONUTRIENTS
芭田 BaTian
Mosaic
PURSELL AGRI-TECH SINCE 1904
UltraYield MICRONUTRIENTS
NutriAg
Compass Minerals
Coromandel
DAYAL
Forbon
IFFCO
iSKY 埃索凯
JG CHEMICALS
江苏华昌化工股份有限公司
JIANGSU HUACHANG CHEMICAL CO.,LTD.
Kingenta International 金正大国际
Matellus
agrichem
微补 Microfeed
chakradhar CHEMICALS PVT. LTD.
OCP
bohigh ENTERPRISE
RioTinto
PRODUQUÍMICA
RUBAMIN
XHF
Anya Polytech & Fertilizers Pvt. Ltd. U.P. INDIA
TOROS AGRI
TATA
YARA
史丹利 STANLEY
TIGER-SUL

CONCLUSION

- ZNI program has had a significant impact in growing zinc fertilizer market

Zinc Nutrient Initiative

- Huge potential remains

- Ongoing activities expected to increase market even further

International Zinc Association

Recycling a small battery, making a BIG Impact; An overview of Alkaline battery recycling

Shane Thompson,
VP-Corporate Development
Battery Solutions

Battery Solutions

February, 13th 2018

Battery Solutions

Battery Solutions

- Founded in 1992 in Southeastern Michigan
- 100+ employees
- Acquired in 2012 by private equity investors with principal investments focusing on clean technology and environmental practices

Alcentra

A subsidiary of BNY MELLON

Centerfield Capital PARTNERS

ALSTON CAPITAL PARTNERS

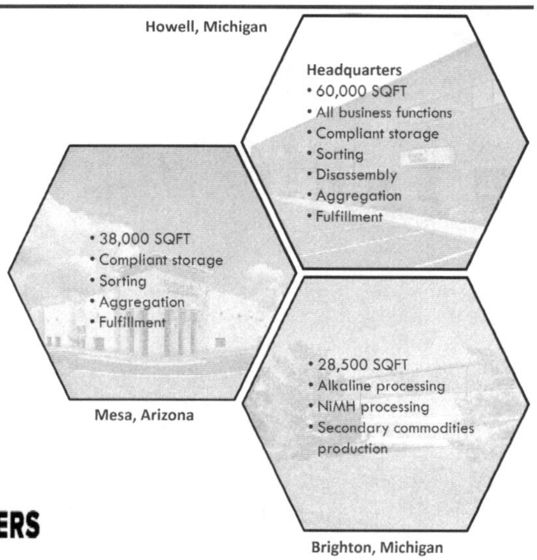

2

Battery Solutions

Battery Solutions

- Largest processor of batteries in North America, processing approximately 25,000 tonnes annually
- First and only North American company to deploy automatic sorting
- First North American company to utilize mechanical alkaline battery processing
- Awarded Highly Commended in The Circulars 2016 by the The Young Global Leaders of the World Economic Forum
- First R2:2013/RIOS™ certified battery recycler

3

Building Blocks to World Class Operations

Battery Solutions

People

Employee training and development programs, employee engagement

Process

Lean manufacturing practices, data tracking and analysis

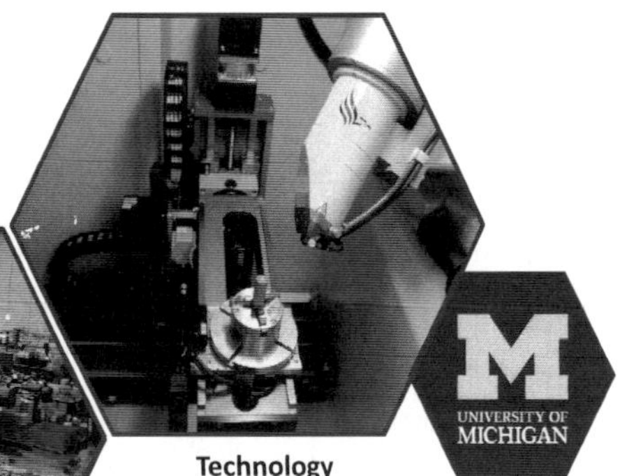

Technology
Research
Innovation

Commitment to industry transforming solutions through innovative use of technology

4

Development of Intelligent Sorting

Battery Solutions

- Requires efficiency to handle sheer volume
- In addition to achieving efficiencies, strive to make sorting a value add activity in itself, sorting by:
  - Type of battery
  - Chemistry
  - Brandmolecular-bond-2
  - Sub-chemistry
  - Rechargeable batteries – remaining life
  - Commercial demand for certain types of batteries
- Allows us to create value in new and innov ways

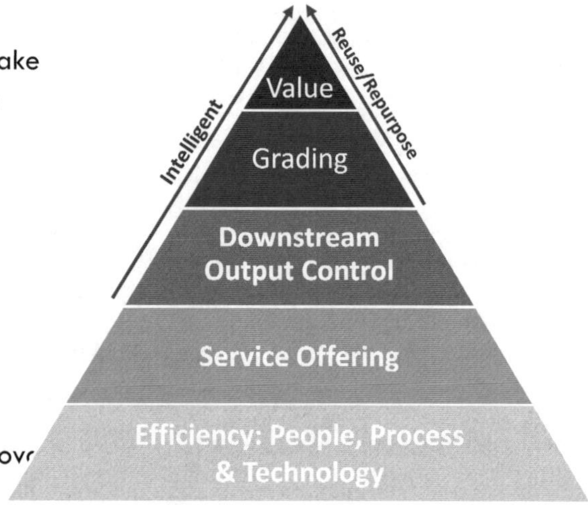

Holistic Approach to Sorting

5

Controlling Downstream Feedstock

Battery Solutions

Our primary research concludes **unbranded** batteries contain substantially higher impurities compared to brand name batteries:

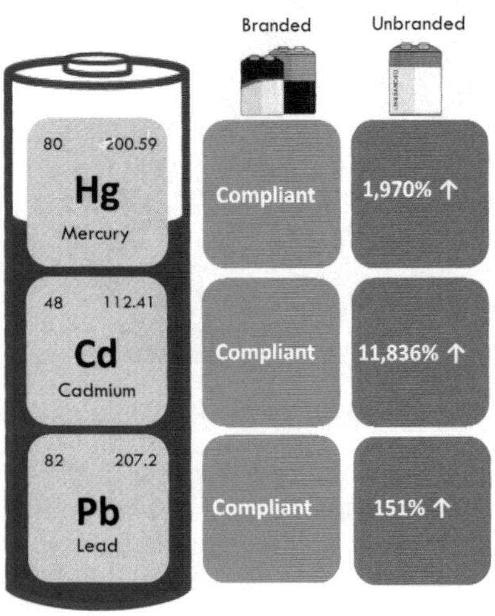

- Depending on the application the quality and purity of the zinc manganese concentrate (ZMC) is more critical
- If we know the application ahead of time, we can sort the batteries to meet the specific criteria

Holistic Approach to Sorting

6

Intelligent Sorting-Capturing Maximum Value

Battery Solutions

The ultimate form of recycling is reuse

- Captures the highest environmental benefit and the best economic value
- A reusable Li-ion cell is on average worth 25x its recycled value

What batteries are worth investing in?

- The aggregate represents enormous value, but each individual cell is not that valuable

Every battery has a story to tell

- Product and usage data – Brand, type, age, application, source
- Test and verification data – Discharge test, X-ray/CT Scan

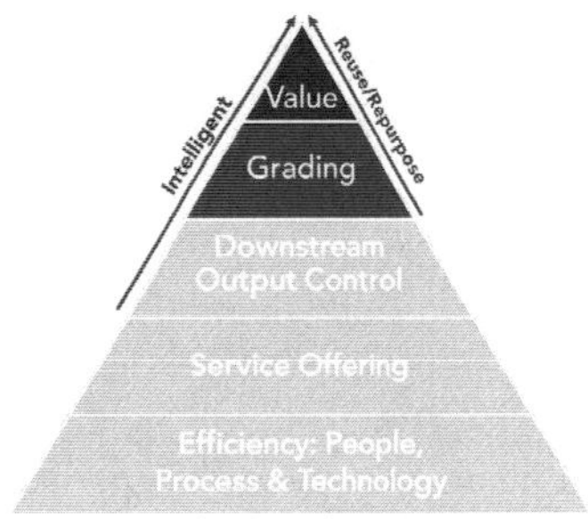

Holistic Approach to Sorting

Use of intelligent sorting technology to achieve very high yield fractions before manual labor is used

- Capable to sort based on commercial, market demand criteria
- Capable to perform automatic grading
- Regulatory implications of sorting data

Building very large data volumes will allow for effective use of machine learning algorithms

7

Sorting to Commercial Criteria and Opti-Sort

Battery Solutions

- Data retrieved from e-commerce engine (tells the sorting machine "What to look for...")
- International network of distributors and resellers providing data (creates a global market for the batteries)
- In real time distinguish for any battery:
  - Worldwide market demand
  - Pricing

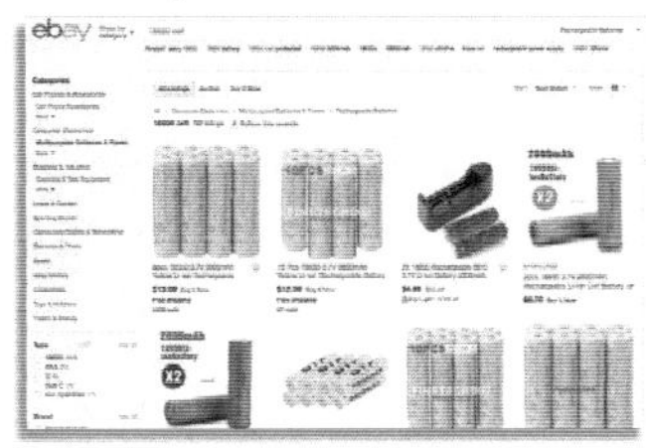

Opti-Sort-great tool for alkaline-zinc carbon batteries-because of the large volume of batteries that need to be handled

8

Automatic Sorting and Grading

Battery Solutions

Design Concept: Integrated Li-Ion Battery Sorter & Grader

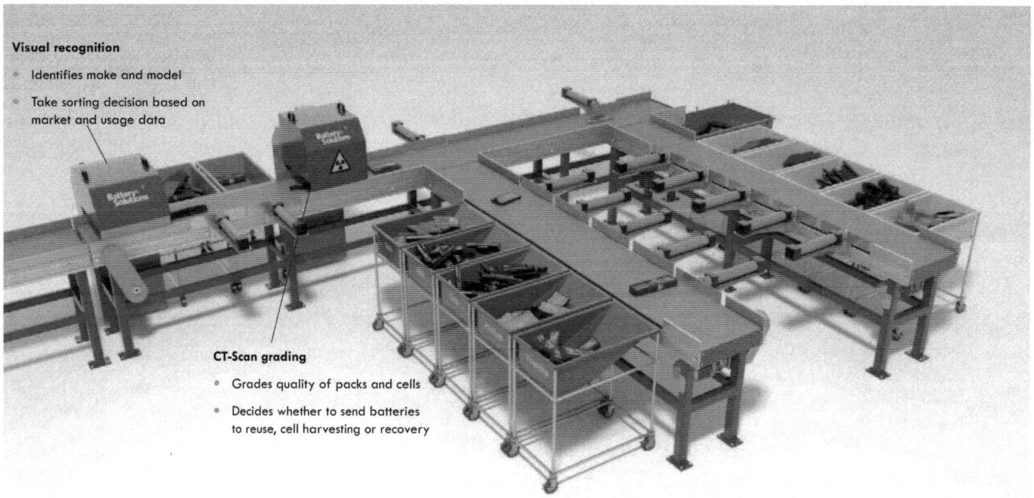

9

US Market Overview

Battery Solutions

- World's largest consumer market
- Few, large national battery collectors
- Resulting in large concentration of volume
- Legal to landfill alkaline batteries in most states, yielding low collection rates for primary batteries

10

US-State Battery Recycling Laws

Battery Solutions

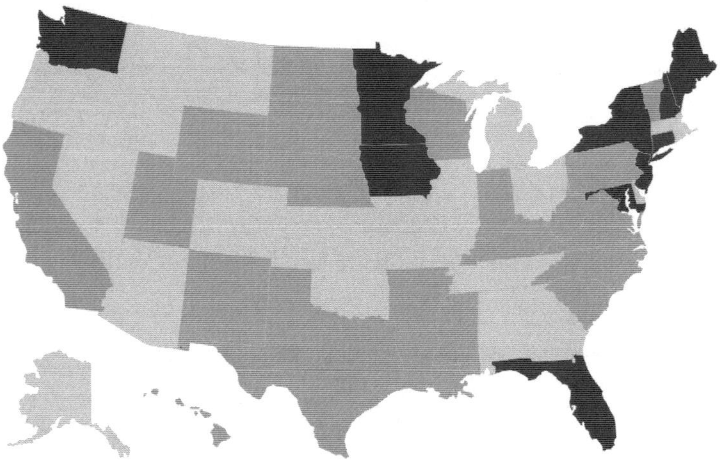

2

10

19

19

States that have battery recycling laws

38

versus

19

Don't *yet* have state battery recycling laws

State battery recycling laws for all chemistries, including alkaline

State battery recycling law for a multiple chemistries

State battery recycling laws for a single chemistry

No state battery recycling law

All states must comply with the 1996 Battery Management Act and The Universal Waste Rule prohibiting disposing of lead, mercury and cadmium containing rechargeable batteries and that these batteries must be recycled.

11

US Market Overview

Battery Solutions

Consumer Batteries Available to Recycle in the US - Excluding Embedded Batteries (kg)

Chemistry	2014	2020	% of total in 2014	% of total in 2020
Alkaline	169,699,444	152,870,234	65.8%	65.5%
Zinc Carbon	11,206,389	7,215,362	4.3%	3.1%
Lithium Primary	1,256,333	1,235,849	0.5%	0.5%
Other Primary	5,942,983	5,169,890	2.3%	2.2%
Li-ion	12,545,397	12,080,266	4.9%	5.2%
NiMH	1,877,357	2,080,174	0.7%	0.9%
NiCd	14,128,640	9,757,675	5.5%	4.2%
SSLA/Pb	41,191,826	42,839,396	16.0%	18.4%
Total Primary	**188,105,149**	**166,491,334**	**73.0%**	**71.4%**
Total Rechargeable	**69,743,220**	**66,757,511**	**27.0%**	**28.6%**
Total Batteries	**257,848,368**	**233,248,846**	**100.0%**	**100.0%**

- US has higher share of primary batteries than most international markets
- Market shifting towards rechargeable
- Alkaline and Li-ion batteries shares are understated when measured in weight
  - Li-ion 38% of weight vs. 83% of units (of sold rechargeable)

Source: Call2Recycle, US Consumer Battery Sales & Available for Collection 2014 to 2020, May 2016

12

Alkaline Processing

Battery Solutions

- Mechanical process at room temperature
- Shreds and hammers to yield output fractions
- Uses magnets and sifting screens for output separation
- Low emissions compared to HTMR

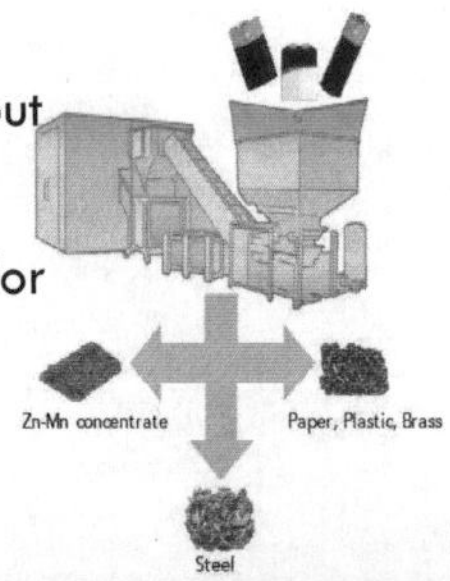

- 97.98% recycling efficiency rate (RER)
- Processes about 6 million pounds annually
- All key equipment parts are sourced in the U.S.
- All three secondary commodities are put back into the marketplace

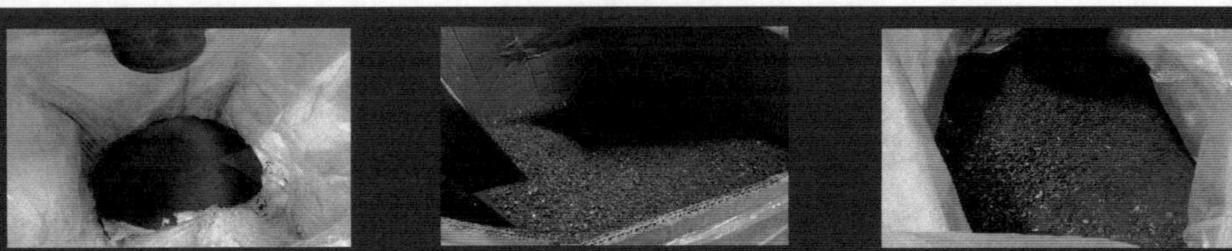

Zinc Manganese Concentrate | Steel | Paper-Plastic-Brass

Optimizing Downstream Recovery Value

Battery Solutions

Examples of ZMC applications

Batteries

Largest segment for Battery Solutions

Micronutrient

Industrial Applications

Brick Colorant
Ferro Manganese

Controlling the alkaline feed stock allows for more effective control of ZMC output for a given application

Micronutrient Fertilizer Market

Battery Solutions

- Excellent feedstock material for production of oxy-sulfate, granular fertilizer's.
  - 90 million acres of corn that need ZnO
  - 85 million acres of soybean that need MnO
- Participated in European LCA focused on use in micronutrient applications
- The majority of (77%) of Zinc Sulfate is used in fertilizer; 23% is used in animal feed
- The majority of (60%) of Zinc Oxide used in animal feed; 40% used in fertilizer

CAMERON MICRONUTRIENTS

AMP™ Advanced Micronutrient Products

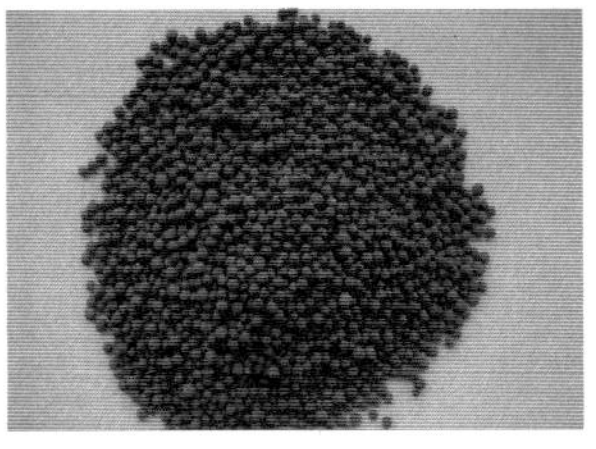

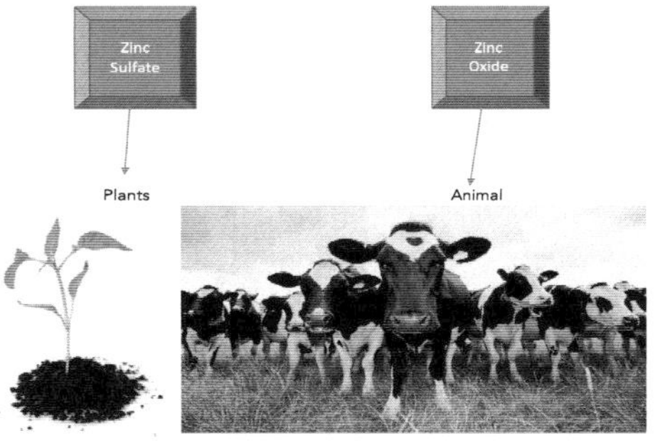

15

Market Overview – Macroeconomics

Battery Solutions

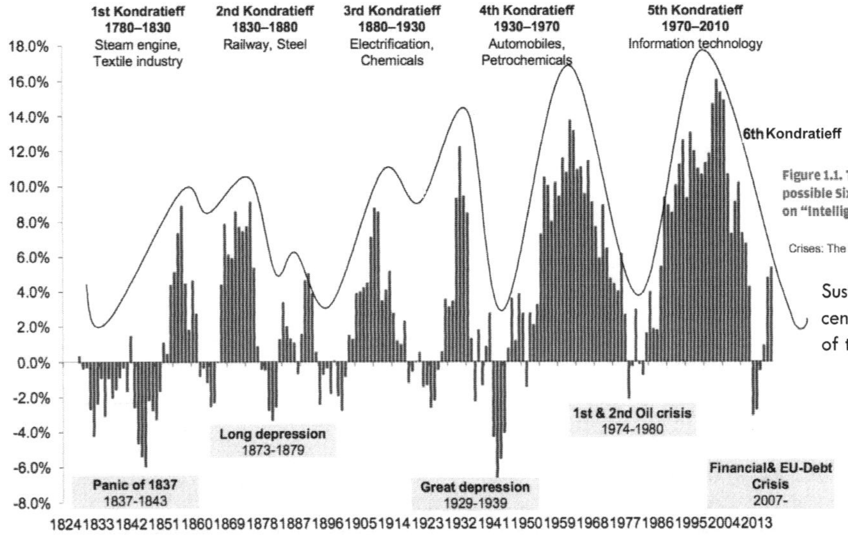

Figure 1.1. The Five Waves of Technology, and the possible Sixth Wave of Sustainable Development built on "Intelligent Technologies" (Allianz)

Crises: The Creative Power of Destruction

Sustainable Development is at the center of the emerging sixth wave of technology

Source: Robert J. Shiller, Stock Market Data used in „Irrational Exuberance", Princeton University Press, 2005, Datastream; Allianz GI Global Capital Markets & Thematic Research

16

Market Overview – Moving Towards a Circular Economy

Battery Solutions

Population: 9 billion by 2030

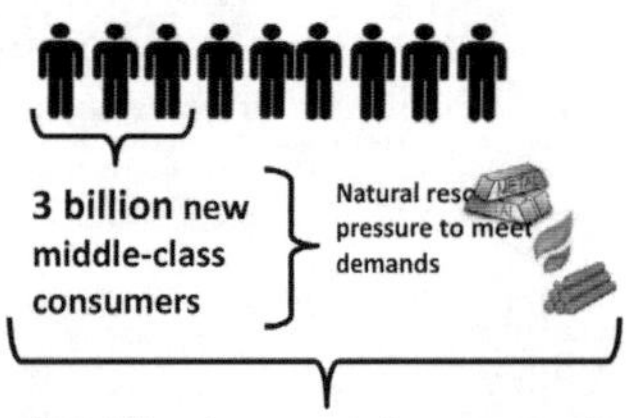

~$3 trillion in economic opportunity through the circular economy model

Global Partners in The Circular Economy

Unilever H&M CISCO PHILIPS RENAULT Google

How Many Earth to Sustain Our Current Lifestyles:

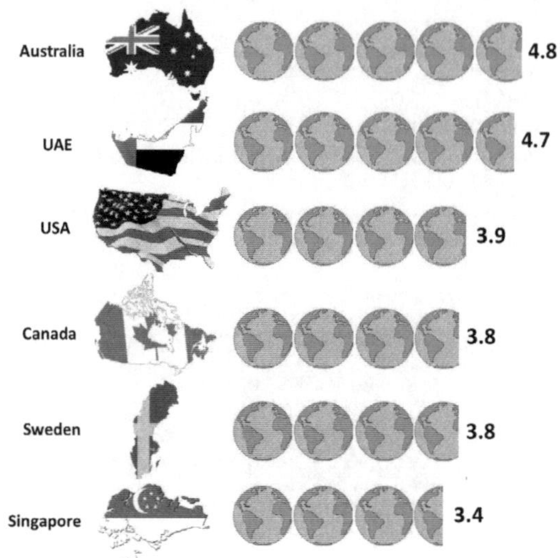

17

The Problem-Zn deficits-Public Health and Economics

Battery Solutions

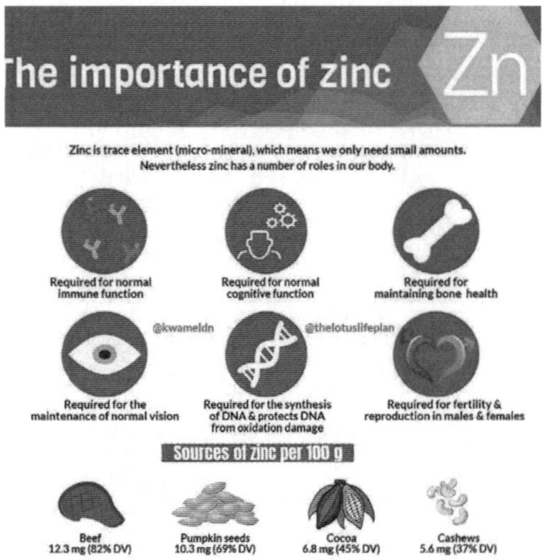

COPENHAGEN CONSENSUS CENTER

- The *Copenhagen Consensus* is a think tank that researches the smartest solutions for the world's biggest problems, advising policy makers and philanthropists on how to spend their money most effectively. The Consensus has Identified Zinc deficiency as a fixable health problem through the use of Zn based fertilizer's.
- They identified 2.2 Billion people who suffer from Zinc deficiency globally, this results in 800,000 deaths/year half of which are children under the age of 5

18

Zn Demand for the Micronutrient Fertilizer Market

Battery Solutions

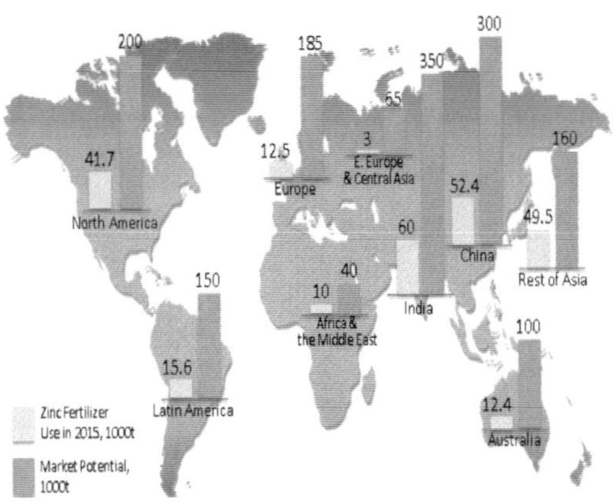

Figure 1: The Cycle of Zinc

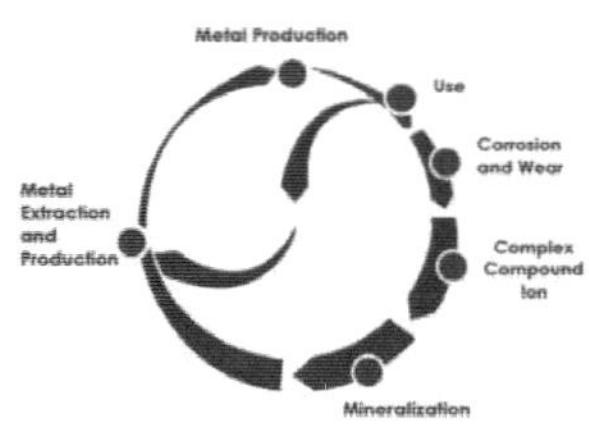

Alkaline Batteries are well suited to be incorporated into the Cycle of Zinc noted above ...

The other graph shows that more Zn will be needed, that market potential is tremendous

19

Agronomic Biofortificaiton

Battery Solutions

- Agronomic Biofortificaiton
  - **Increases crop yields** and...
  - **Provides more dietary Zinc...a literal "Win-Win"**
- International Zinc Association states that it is estimated that in the next 50 years farmers will have to produce more food than was produced in the last 10,000 years COMBINED.
- Will have to come from increasing production on existing farmland (Higher crop yields, sustainable lands)
- Need for Zinc in contributing to food and nutritional security by:
  - Providing key micronutrients for the growth and health of all plants
  - Increasing crop yields
  - Addresses Zn defiency, both in terms of public health as well as in economic growth
  - Contributes to agricultural productivity by getting more OUT of the land rather than get more by INCREASING acres.

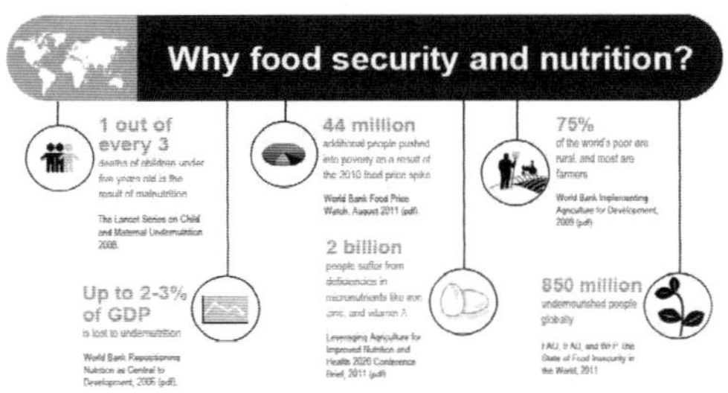

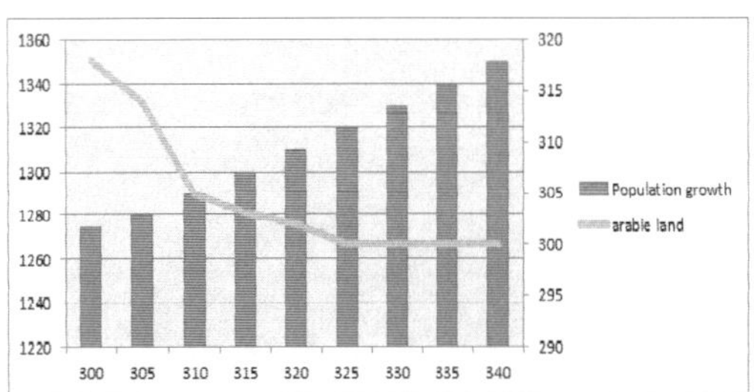

20

ZnO Potential

Battery Solutions

Consumer Batteries Available to Recycle in the US - Excluding Embedded Batteries (kg)

Chemistry	2014	2020	% of total in 2014	% of total in 2020
Alkaline	169,699,444	152,870,234	65.8%	65.5%
Zinc Carbon	11,206,389	7,215,362	4.3%	3.1%
Lithium Primary	[illegible]	[illegible]	0.5%	0.5%
Other Primary	42,983	5,169,890	2.3%	2.2%
Li-ion	5,397	12,080,266	4.9%	5.2%
NiMH	,877,357	2,080,174	0.7%	0.9%
NiCd	4,128,640	9,757,675	5.5%	4.2%
SSLA/Pb	41,191,826	42,839,396	16.0%	18.4%
Total Primary	**188,105,149**	**166,491,334**	**73.0%**	**71.4%**
Total Rechargeabl	**69,743,220**	**66,757,511**	**27.0%**	**28.6%**
Total Batteries	**257,848,368**	**233,248,846**	**100.0%**	**100.0%**

- Alkaline at 25% ZnO recovery = 42,039 tons of ZnO
- Zinc Carbon at 20% ZnO recovery =1,587 tons of ZnO
- Could add Zn from additional batteries; Nickel Zinc, Zinc Air; both emerging large format type batteries Could contribute additional tons of Zn or ZnO units per year

Source: Call2Recycle, US Consumer Battery Sales & Available for Collection 2014 to 2020, May 2016

21

Thank you

Shane Thompson, VP
Battery Solutions

Battery Solutions

February 14, 2018

U.S. TIRE MANUFACTURERS ASSOCIATION

Tracking US Tire Shipments and Regulatory Trends in Tire Materials

February 14, 2018

USTMA Member Companies

BRIDGESTONE
Hankook driving emotion
SUMITOMO RUBBER INDUSTRIES
Continental
KUMHO TIRE Better, All-Ways
TOYO TIRES driven to perform
COOPERTIRES
MICHELIN A BETTER WAY FORWARD
YOKOHAMA
GOODYEAR MORE DRIVEN
PIRELLI

U.S. TIRE MANUFACTURERS ASSOCIATION

US Tire Manufacturing Industry

- $148.4 billion dollar industry
- Responsible for more than 737,000 jobs in the U.S.
  - Directly supports more than 284,000 U.S. jobs through manufacturing, distribution and retailing
  - responsible for more than 450,000 additional U.S. jobs in supplier and induced activities

U.S. TIRE MANUFACTURERS ASSOCIATION 3

What we will cover

- Tracking US Tire Industry Shipments
- Changes to the Toxic Substances Control Act
- Crumb Rubber Studies
- California concerns about zinc in stormwater

U.S. TIRE MANUFACTURERS ASSOCIATION

TRACKING US TIRE INDUSTRY SHIPMENTS

USTMA Data Program

- Tracking the US Tire Industry Shipments:
    - Shipment data is reported by USTMA member companies which represent about 85% of the US tire industry
    - Estimate non-member activities by analyzing total US imports and USTMA member imports
    - Covers Original Equipment (OE), Replacement, Production, Imports, Exports markets for passenger, light truck and commercial truck tires.
    - Conducts industry shipment forecasts, three time a year
    - Represents the primary and sole source of information prepared by and for the US tire industry

US Tire Market trends (OE & Replacement)

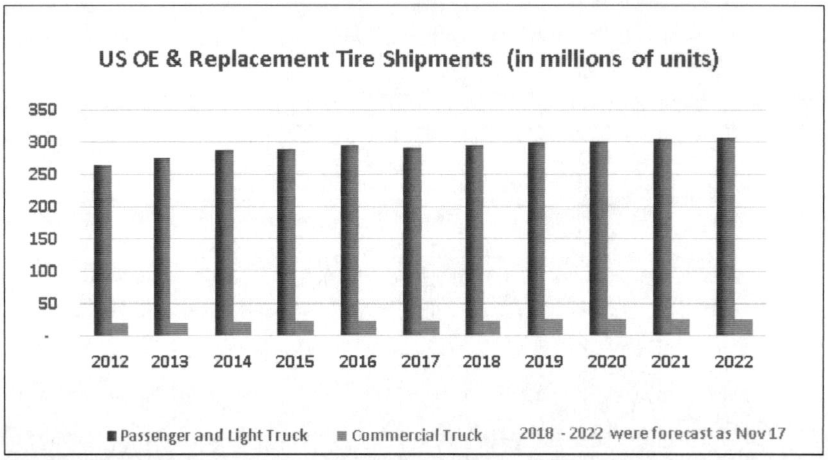

USTMA Data Publications

- USTMA publishes three tire market data reports:
  - Monthly Tire Shipment Report
  - Annual Tire Industry Factbook
  - Monthly US Tire Trade Report
- Questions regarding the information presented in these publications should be directed to the Market Information Services department of USTMA at 202-682-4865
- These USTMA data publications are available for purchase on our website: www.ustires.org/publications

8

USTMA Monthly Tire Shipment Report

U.S. TIRE INDUSTRY SHIPMENT SUMMARY *(thousands of units)*

	Original Equipment	*Replacement*	*Exports (USTMA)*	*Total Shipments*	*Production (USTMA)*	*US Imports*
	Total	Total	Total	Total	Total	Total
Jan	4,696	17,268	2,781	24,745	13,983	13,228
Feb	4,864	16,090	2,658	23,612	13,128	10,566
Mar	5,494	21,569	2,902	29,965	14,996	14,341
Apr	5,245	22,209	2,663	30,118	14,161	18,102
May	5,126	21,145	2,736	29,007	14,570	17,059
Jun	5,284	23,822	3,160	32,266	14,043	18,261
Jul	4,615	22,955	2,985	30,554	12,637	18,997
Aug	5,288	20,225	2,761	28,275	14,950	15,254
Sep	5,243	21,699	3,056	29,997	14,281	14,961
Oct	5,369	22,029	3,012	30,410	15,016	14,889
Nov	4,848	22,805	2,641	30,295	13,655	15,892
Dec	4,211	21,381	2,312	27,904	11,987	15,101
Total	60,284	253,198	33,667	347,149	167,407	186,650
Month % Change For December						
2015/2014	-0.1%	11.4%	-5.4%	7.9%	0.9%	15.6%
YTD % Change Through December						
2015	60,284	253,198	33,667	347,149	167,407	186,650
2014	58,104	252,606	32,871	343,581	161,594	188,048
%	3.8%	0.2%	2.4%	1.0%	3.6%	-0.7%

U.S. TIRE MANUFACTURERS ASSOCIATION 9

USTMA Monthly Tire Shipment Report

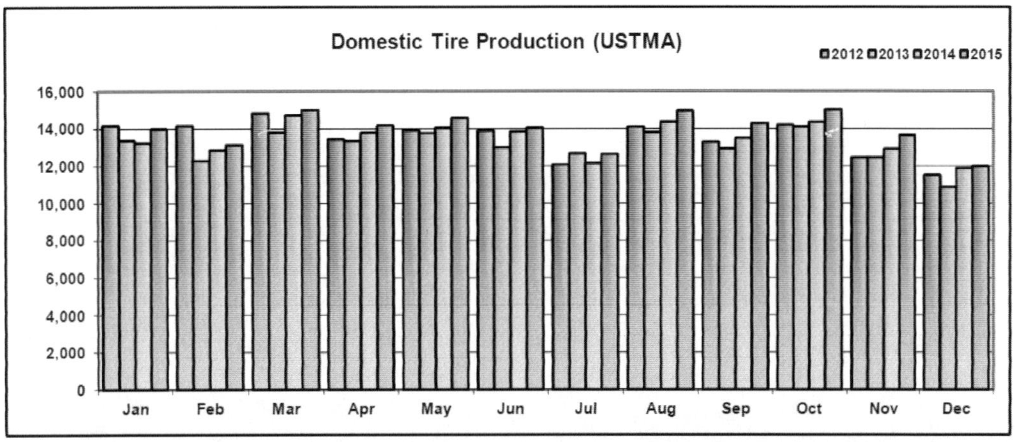

U.S. TIRE MANUFACTURERS ASSOCIATION 10

REGULATORY TRENDS IN TIRE MATERIALS

TOXIC SUBSTANCES CONTROL ACT (TSCA)

Impacts of TSCA reform

- Greater focus on materials in products
- Contaminants are not excluded from review under TSCA

U.S. TIRE MANUFACTURERS ASSOCIATION

13

TSCA History

October 11, 1976 – TSCA, Public Law 94-469, signed into Law

40 years later…

June 22, 2016 – The Frank R. Lautenberg Chemical Safety for the 21st Century Act, Public Law 114-182, signed into law (bipartisan legislation)

- Effective immediately, started clock on implementation

U.S. TIRE MANUFACTURERS ASSOCIATION

Key Changes

- **Reset of the TSCA Inventory**: Active and Inactive
- **Prioritization of Active Chemicals**: Low Priority and High Priority
- **Risk Evaluations** on High Priority chemicals, may result in use restrictions/risk management
- **Order authority**: EPA can order the generation of data and the performance of testing
- **Fees**: EPA can charge fees at various points in the regulatory process

U.S. TIRE MANUFACTURERS ASSOCIATION

Process for reviewing substances under TSCA

Prioritization
- Chemical designated as a low priority
- Chemical designated as a high priority

Risk Evaluation
- No unreasonable risk
- Unreasonable risk

Risk Management
- Impose restrictions to eliminate unreasonable risk

U.S. TIRE MANUFACTURERS ASSOCIATION

Prioritization

- Process EPA will use to identify potential candidate chemicals for EPA's prioritization process under TSCA
- EPA public meeting (December 11, 2017) - Approaches for Identifying Potential Candidates for Prioritization for Risk Evaluation Under Amended TSCA
- EPA public comment period on 6 ideas for prioritization – closed January 25th

U.S. TIRE MANUFACTURERS ASSOCIATION 17

Prioritization – USTMA Advocacy

- Downstream user coalition - Plastics, Toys, Auto Parts, Paper and Wood Products, and Tires
- Support implementation of the amendments to TSCA
- Support a robust federal approach to the regulation of chemicals
- Support a process that provides any early understanding of which chemicals EPA will review

U.S. TIRE MANUFACTURERS ASSOCIATION 18

Prioritization – USTMA Advocacy

- Establish a permanent and predictable "Information Outreach Framework"
  - Process to facilitate the exchange of information by stakeholders all along the value chain
  - Downstream users – important resource for information on conditions of use
  - Mechanism for companies to share accurate use and exposure data, correct errors and supplement with newly discovered information within a reasonable period of time

U.S. TIRE MANUFACTURERS ASSOCIATION 19

Implementation – Where are things now

- TSCA Inventory Reset
- First 10 chemicals for risk evaluation
- Prioritization of additional substances – Could zinc oxide be prioritized?
- Fees rule

U.S. TIRE MANUFACTURERS ASSOCIATION

FEDERAL AND STATE CRUMB RUBBER STUDIES

21

Crumb Rubber

National Toxicology Program (NTP)
- End of 1st/ beginning of 2nd quarter 2018

U.S. Federal Study (ATSDR, CPSC, EPA)
- Early 2018?

California (OEHHA)
- May 2019

22

CALIFORNIA CONCERNS WITH ZINC IN STORMWATER

California and zinc

- Zinc oxide is an essential ingredient in manufacturing tires
- State of knowledge – to better understand sources in zinc in stormwater

California and zinc

- Updated calculations on tire wear and zinc that can leach from tire wear
- Shared information with CA Water Board with support of IZA – removal of zinc from STORMS
- Continue to support a statewide variance for zinc

U.S. TIRE MANUFACTURERS ASSOCIATION

THANK YOU

SARAH E. AMICK
U.S. TIRE MANUFACTURERS ASSOCIATION
VICE PRESIDENT EHS&S AND SENIOR COUNSEL
SAMICK@USTIRES.ORG

U.S. TIRE MANUFACTURERS ASSOCIATION

26

Reduce Production Costs and Increase Productivity using Energy Efficient Crucibles

Mike Zeno
Vesuvius-Foseco

Vesuvius-Foseco
Crucible Technology
A Move Toward Energy Efficiency
Examples of Success
Challenges
Actions to Improve Your Bottom Line

VESUVIUS

FOSECO

VESUVIUS

FOSECO

- VESUVIUS (Cookson, Founded 1704)
  - Metals and Glass business
  - 1987, Vesuvius Crucible Company
  - 2008, Acquired FOSECO
- FOSECO (Founded 1936)
  - Foundry Service Company
  - Foundries, Steel Mills, Refineries
- 10,000 employees, 70 manufacturing plants, in 40 countries

FOSECO

- A Global Leader, Engineered Clay Graphite and Silicon Carbide crucibles. State-of-the-art R&D Facility
- Four global production sites, 6,000+ MT of crucibles annually for a wide variety of industries and applications
- Supplier of crucibles for Zinc Oxide production

Crucible Technology

Crucible Technology

- Earliest Crucible forms
  - Made from clay or stone, “Pots”
  - Poor Thermal Conductivity
  - High energy to melt
- Clay Graphite / Silicon Carbide
  - Better Efficiency / Thermal Conductivity
  - Versatile, Durable, Thermally Stable
- Engineered Shapes / Compositions
  - Designed to Melt Specific Materials
  - Durable, Thermally Stable
  - Highly Efficient

A Move Toward Energy Efficiency

A Move Toward Energy Efficiency

Reducing cost is critical to staying competitive in the market

1. Utilize a low-wage workforce
   - (High Turnover, unsustainable)
   - inexperience = costly mistakes
2. Purchase cheaper consumables
   - (risky) early failure = more costly
   - Insignificant relative to other Op.Costs
3. Reduce energy costs
   - (can be difficult, variable but effective)
4. Increase productivity
   - (can be difficult, but effective)

A Move Toward Energy Efficiency

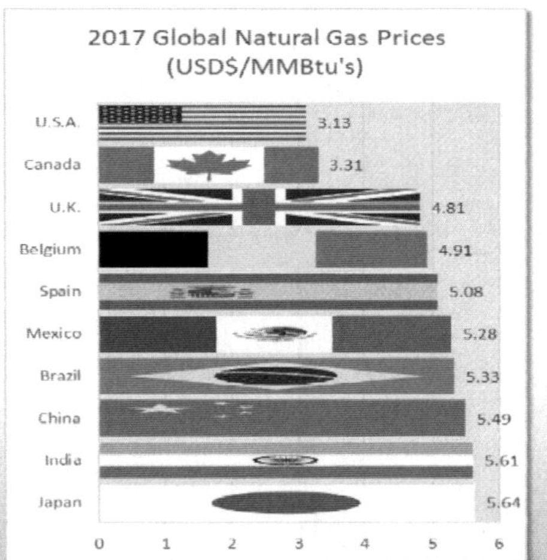

- Natural Gas, typically used for Zinc Oxide production
  - A significant cost variable
  - Varies by Geographic location
- Typically 8-10 Million BTU (or GJ) energy per MT ZnO produced
- Demand of 10 GJ / 10 Million BTU's / 250 m3 Natural Gas in North America means approx. **US $30 per MT Zinc Oxide**
- By comparison, crucible cost is ~$2-3 per MT ZnO

https://www.statista.com/statistics/252984/landed-prices-of-liquefied-natural-gas-in-selected-regions-worldwide/

A Move Toward Energy Efficiency

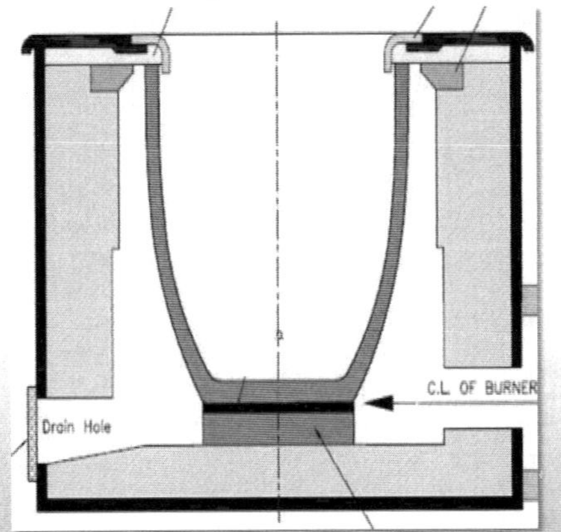

- Vaporizing Zinc in flame-fired furnaces, Energy intensive
  - Furnace Efficiency
  - Crucible Thermal Efficiency / Conductivity
- Furnace efficiency itself varies based on design and materials.
  - Furnace Replacement/Refurbish can be costly, materials & downtime
- Improved Crucible Thermal Efficiency directly impacts:
  - Energy consumption
  - Zinc Oxide output (Productivity)

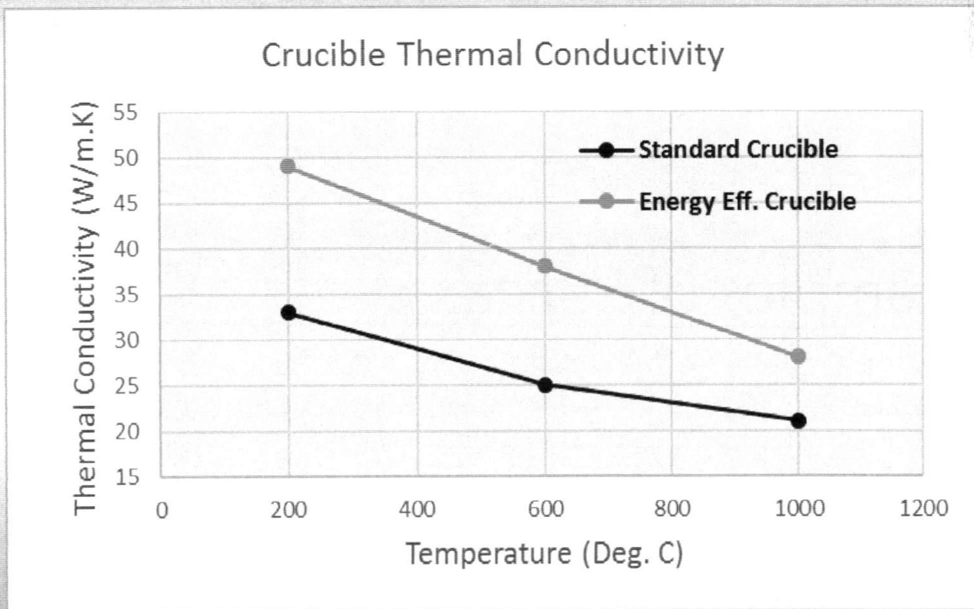

A Move Toward Energy Efficiency

ENERTEK ZnO

- Ideally suited to Indirect or French Process
- Extremely high thermal conductivity combined with high oxidation resistance
- Faster production cycles
- Significantly less energy use per MT Zinc Oxide produced
- Reduced emissions and CO2 generation (carbon footprint)
- Lower overall production costs per MT Zinc Oxide

Examples of Success

Example:

- Zinc Oxide Output equal
- ENERTEK ZnO crucible 10% greater price
- Crucible life (90 days).
- ~ 20% energy savings with ENERTEK ZnO
- +400GBP per year = Net Savings of 17K GBP/yr or 7.8GBP per MT ZnO

Natural Gas Furnace		Competitor	ENERTEK ZnO	% Difference
Production	Rated Period (Days)	90	90	0%
Zinc Oxide Output	Output (MT) per day, per crucible	6.0	6.0	0%
	Output (MT) for Rated Period	540	540	0%
Crucible	Cost (GBP)	1,000	1,100	10%
	Life (Days)	90	90	0%
	Cost for Rated Period (GBP)	1,000	1,100	10%
Energy	Use per Day (m3)	800	640	-20%
	Use per MT ZnO output (m3)	133	107	-20%
	Used Over Rating Period (m3)	72,000	57,600	-20%
	Cost (GBP/m3)	0.30	0.30	0%
	Cost for Rated Period (GBP)	21,600	17,280	-20%
Process Costs in Rated Period	Total Crucible + Energy (GBP)	22,600	18,380	-19%
	Total Cost (GBP) per MT ZnO	41.9	34.0	-19%
NET Savings with ENERTEK ZnO Crucibles	Estimated Annualized (GBP)	17,114		
	per MT ZnO (GBP / MT)	7.8		

SCENARIO A - Value of Energy Savings alone

Customer reported 33% energy improvement when using ENERTEK ZnO crucibles
No potential for increased business - Annual Zinc Oxide output of 3,600 MT remains unchanged.

- Customer uses 8 crucibles per year (450 MT ZnO per crucible).
- ENERTEK ZnO priced **€700** greater than standard crucible **(€5,600 more per year)**.

Assessment of Energy Saving only (Annual Savings)

- Current gas usage @ 396,000m^3 (110m^3 gas per MT ZnO produced)
- 33% gas reduction = 132,000m^3
- Gas price = €0.40 / m^3
- Annual gas saving = **€53,000**

- Net Customer Benefit of **€47,400 per year**
- **Reduction in CO_2 emissions of 284 tonnes was achieved** (2.15kg CO_2 per m^3 gas consumed).

SCENARIO B - Value of Increased Productivity alone

Customer reported a 33% production rate improvement when using ENERTEK ZnO crucibles
Increased productivity leveraged by increasing annual Zinc Oxide output from 3,600 to 4,800 MT

Assessment - based on increased production output

- Additional 1,200 MT Zinc Oxide per year made and sold **without any capital investment and same energy consumption**.

- Additional sales revenue of **€1.8m from this increased output** (estimated ZnO market price ~ €1,500 per MT).
- Estimated additional profit > **€400,000** (revenue after deducting ZnO variable manufacturing costs, incl. additional crucibles).
- **Crucible cost is insignificant relative to increased revenue**

Examples of Success

Customer	ZnO Producer, Europe	ZnO Producer, Europe	Zinc Dust Producer, Europe
Crucible Tested	ENERTEK ZnO Retort	ENERTEK ZnO Retort	ENERTEK ZnO BP Type
Objective	Reduce Energy, Increase Productivity	Reduce Energy, Increase Productivity	Reduce Energy, Increase Productivity
Alloy	Zinc	Zinc	High Purity Zinc
Benefit 1	30% Higher Productivity	13% Higher Productivity	55% Higher Productivity
Benefit 2	NG reduced 132,000 M^3/yr	Oil use reduced 12%	NG use reduced 34%
Benefit 3	266 MT Less CO^2 per yr	224 MT Less CO2 per yr	34% Less CO2 per yr
Benefit 4	Annual savings = €53,000	Annual savings = €18.5K	25% Longer Crucible Life

Challenges Faced

- Crucibles often considered as commodities where only purchase price is important. Not True !!
- Some furnaces designed to contain multiple crucibles.
- Fuel consumption not always measured at each furnace
- Fuel use and/or ZnO output not accurately reported, over a given time period.

Improve Your Bottom Line

- <u>Evaluate your current system</u>
  - Gas Meter for every furnace
  - Record fuel usage at each furnace regularly
  - Measure the efficiency and productivity of your current system

- <u>Test energy efficient crucibles</u>
  - Ask for a free sample crucible
  - Measure energy efficiency and productivity of that crucible
  - Compare to the Standard Crucible

<u>Key Takeaway Points</u>

- Reduced energy and faster zinc oxide output = major operating benefits
- Energy costs are the most significant operating costs that can be influenced.
- Crucible cost is relatively minor, but their effects can be major
- Be willing to work with your crucible partners to find the optimal crucible for your operation
- Evaluate your current system, and improve monitoring
- Crucible testing needs to carefully conducted with attention to key process parameters (HINT: Ask suppliers for their assistance) ☆

Mike Zeno

Product Manager - Crucibles

Foseco

20200 Sheldon Rd Cleveland Ohio 44142 USA

Office: 1-440-863-2785

Mobile: 1-440-552-0698

Email: mike.zeno@Foseco.com

Web: https://www.vesuvius.com

LinkedIn: https://www.linkedin.com/company/foseco

Update: EU Regulatory Issues and beyond

Christine Spirlet
Manager, Regulatory Affairs

Overview

- REACH - **R**egistration, **E**valuation, **A**uthorization and Restriction of **Ch**emicals
- Human volunteer study
- Nano regulatory update
- Veterinary medicine, oral care, cosmetics
- What about Pb classification?
- Global issues
- Conclusions

REACH update for ZnO

Zinc REACH Consortium

Consortium membership
- 86 effective members of which 49 registered ZnO
- LoA (Letter of Access) ZnO : 79

Total number of companies registered to date reaching over 130 and still growing

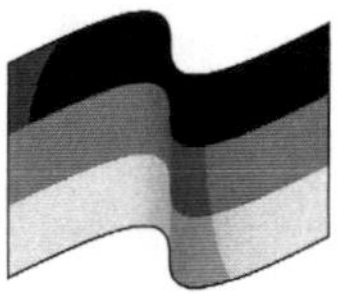

REACH ZnO dossier evaluation - background

Start summer 2015
- August 28th : "draft decision letter" from ECHA
- October 5th Submission comments on the draft decision

→

March 14th 2016
- Informal call with ECHA to discuss comments
- Update of section 1.2 accordingly on April 15th

→

May 30th 2016
- Termination of compliance check procedure on substance identity
- ECHA informs the German Competent Authorities accordingly

Update 2017

April 5th : Meeting in Germany with BAuA, BfR and UBA → Discussion on information present in dossier and their expectations → August 31st : Update of CSR as agreed → October : update of IUCLID → Opinion expected March 21st 2018

Human volunteer study

2018 IZA Conference - Carlsbad (CA)

International Zinc Association

Final IPA report Sept 2017 – conclusions

Most parameters showed no effect at any concentration

Concentration-dependent increase of flu-like symptoms or inflammatory markers was significant at 1.0 mg/m^3 or higher

Inflammatory markers in sputum were increased starting from 0.5 mg/m^3 but without concentration-response relationship.

All observed effects were reversible 14 days after last exposure

Study concludes on a No Observed Effect Level < 1.0 mg/m^3 ZnO

IPA: Institut für Prävention und Arbeitsmedicin (Bochum, DE)

2018 IZA Conference - Carlsbad (CA)

International Zinc Association
8

Revision of the ZnO OEL: way forward

DNEL / OEL between 0.1 and 1 mg/m³

New value: ZnO levels expressed as **respirable fraction, systemic**

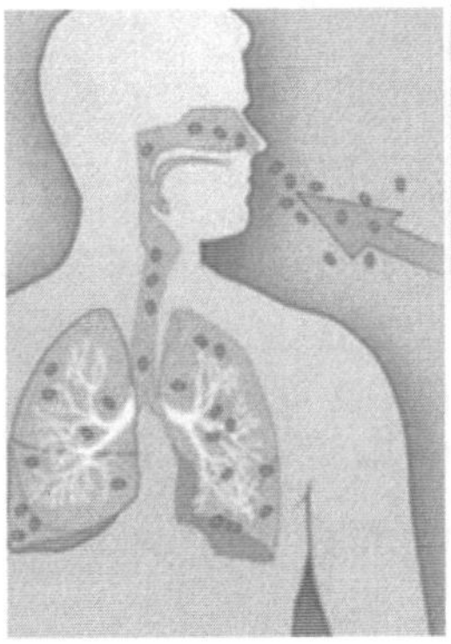

Impact on zinc and other zinc substances

- Slightly solubles (ZnO type): 5 mg Zn/m³
- Fully solubles : 1 mg Zn/m³

Nano regulatory update

Guidance for implementing definition

Roadmap (tentative)		Today?
Draft revised definition	Nov/Dec 2016	Announced Q1 2018
Public consultation	Nov/Dec 2016 – Jan/Feb 2017	Announced following draft publication
Revised definition – adoption	Feb/Mar 2017	Summer?
JRC Report 1 (supporting implementation of definition)	Adoption of the revised definition	Q4 2018
JRC Report 2	End 2017	?

Nano is BIG in EU

https://euon.echa.europa.eu/
Launched June 2017

2nd phase: content development from:
NanoData (information on safety, markets, innovation)
EU funded research projects
National registries
Other EU sources (EFSA,EMA, EU-OSHA, EEA),
Commission services, and other legislations

but not pigs…

ZnO in Veterinary Medicine

2018 IZA Conference - Carlsbad (CA) International Zinc Association

ZnO banned in Veterinary Medicinal Products (EU Comm. June 2017)

History…

Gutal (ZnO) used to treat post-weaning diarrhea in piglets.

FR and NL referred to the Committee for Veterinary Medicinal Products requesting that the marketing authorization be removed for the use of ZnO in veterinary medicinal products (VMP).

The risk for co-selection of antimicrobial resistant bacteria, combined with new (flawed) environmental data = Risk > Benefit.

2018 IZA Conference - Carlsbad (CA) International Zinc Association

ZnO banned in Veterinary Medicinal Products (EU Comm. June 2017)

IZA/ZOPA action taken:

- Scientific level – Flawed environmental data; attended the oral hearing at the European Medicines Agency in Feb. 2016.
- Member State level – Visited a number of countries (UK, IE, PT, ES, DK).

Outcome & Lessons learned:

- MS's have a 5-yr deferral period. Immediate ban feared, deemed a success.
- A political decision from FR and NL? Lack of opposition influenced by the Brexit?
- Marketing authorization holder's acted independently, always an uphill battle...
- Things moved fast, earlier awareness would have been helpful for us.

ZnO in Oral Care

2018 IZA Conference - Carlsbad (CA)
International Zinc Association

Background

Tolerable total uptake level (UL) for Zinc up to 25 mg (2003 SC Food) → Currently allowed for use in oral care products: **1%** as Zinc (2009)

BfR (DE): 1% Zn in toothpaste and mouthwash can lead to excess Zn intake (focus on children and teens) → Feb. 2016, Cosmetics Europe: aggregate exposure assessment for children and adults: **1% toothpaste, 0.1 % for mouthwash safe**

SCCS opinion (March 2017)-first submission

SCCS has estimated that exposure to water-soluble zinc salts via toothpaste at the concentrations of 1% per se is safe.

But SCCS cannot advise which portion of the upper limit should be allocated to exposure from cosmetic products.

- The SCCS is aware that upper limits may be exceeded in some cases because the default values used in this Opinion are based on conservative estimates (e.g. 25mg Zn/day).

Cosmetics Europe has replied to public consultation

Final opinion awaited

Cosmetics

Regulation amendment

Zinc Oxide restriction

- COMMISSION REGULATION (EU) 2017/1413 of 3 August 2017 amending Annex IV to Regulation (EC) No 1223/2009 of the European Parliament and of the Council on cosmetic products

Entry 144 of Annex IV to Regulation (EC) No 1223/2009 is replaced by the following:

Reference number	Substance identification					Conditions			Wording of conditions of use and warnings
	Chemical Name	Colour index Number/Name of Common Ingredients Glossary	CAS number	EC number	Colour	Product type, Body parts	Maximum concentration in ready for use preparation	Other	
a	b	c	d	e	f	g	h	i	j
'144	Zinc oxide (*)	77947	1314-13-2	215-222-5	White			Not to be used in applications that may lead to exposure of the end-user's lungs by inhalation.	

(*) For use as a UV-filter, see Annex VI, No 30 and No 30a.'.

82
Pb
207.2

Updated Pb classification

Pb metal classified

9th ATP

Pb powder classified as Repr. 1A H360D

SCL ≥ 0.03%

What about ZnO?

- In ZnO, Pb is present as PbO
- PbO is classified in item "lead compounds" in CLP with CL ≥ 0.3%
- exposure is to **PbO** and it is the compound that **should be taken into consideration when classifying and assessing the risk**

Global issues

Towards Global Harmonisation?

Since 2008

- IMO, WHO,...
- Consistent information worldwide
- Review of existing (older?) classifications?
- resources needed
- Double work - Legal issues

Pilot project (OECD)

- Comparison EU Annex VI list and Japan GHS list
  ⇒ Of 89 common substances... NONE had identical classification

Decision pending, industry to monitor closely

- UN GHS subcommittee: ICMM
- EU: Eurometaux

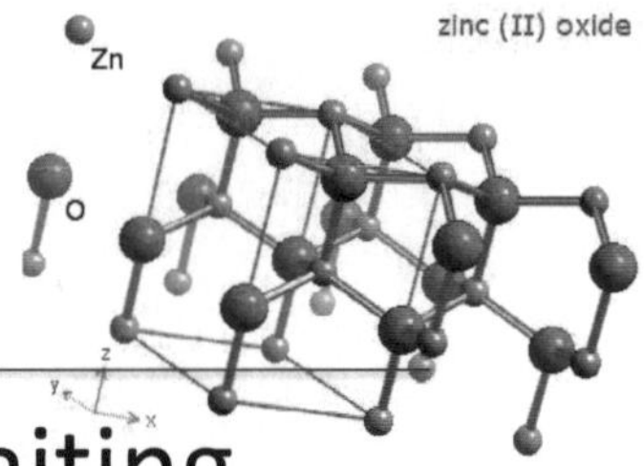

Conclusions

REACH dossier evaluation: waiting

OEL revision: provisional value, t.b.c.

Veterinary Medicine: 5y deferral

Oral Care: inconclusive

Thank you!

Trajectory of Environmental Regulations in the Americas

Eric Van Genderen, PhD
International Zinc Association
14 February 2018

International Zinc Association

Take Away

Canada	Rapid succession of regulatory updates to support CMP Provincial/Territorial sovereignty allows flexibility
USA	Updated and harmonized federal guidelines for water State sovereignty brings uncertainty and flexibility
Latin America	Progressive GHS implementation underway Mosaic of chemical management schemes anticipated

International Zinc Association

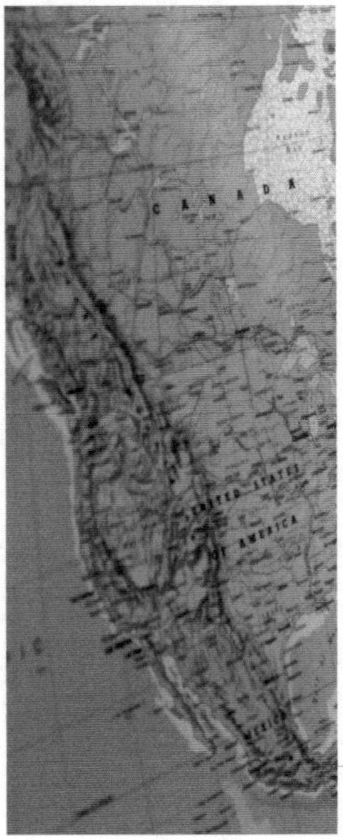

Overview

- Global regulatory perspective
- Canada
  - ➢ CMP, CCME, ECCC/HC
- United States of America
  - ➢ USEPA, CA DTSC
- Latin America
  - ➢ GHS and chemical management

International Zinc Association

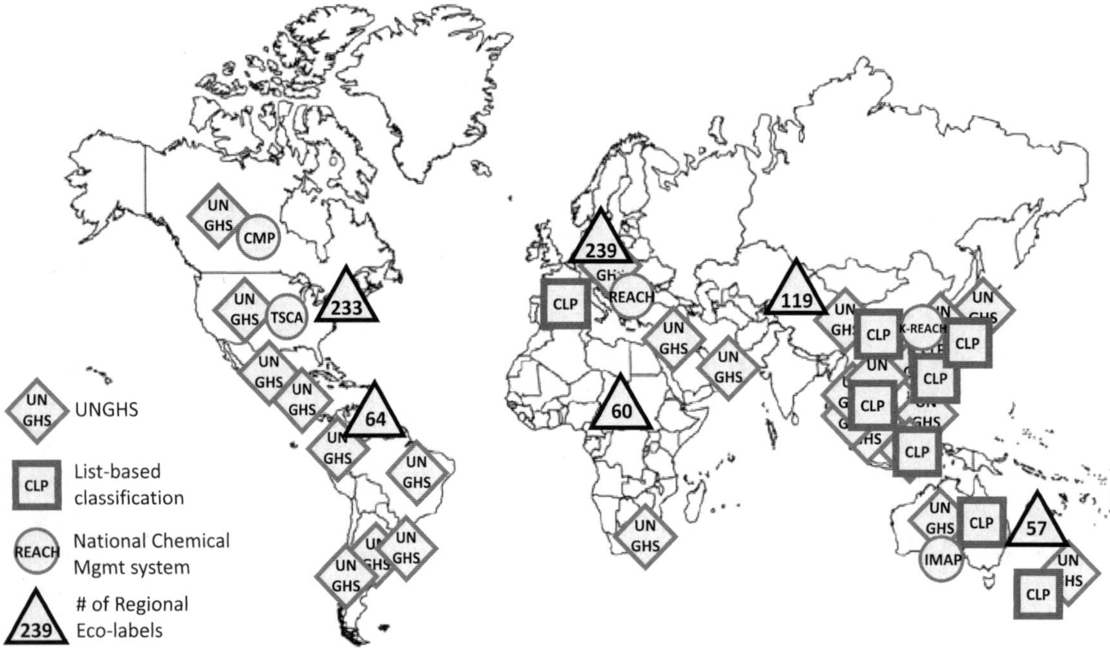

Consequences of Dis-Harmonization

- Market access challenges associated with mosaic of classifications
- Migration of inappropriate GHS classifications into developing economies
- Classifications for substances in consumer products are prohibitive

More like...
Global
Havoc
System

Required GHS Symbol:
Reproductive Toxicity (Cat. 2)
STOT-SE & STOT-RE (Cat. 1 & 2)

Regulatory Life Cycle

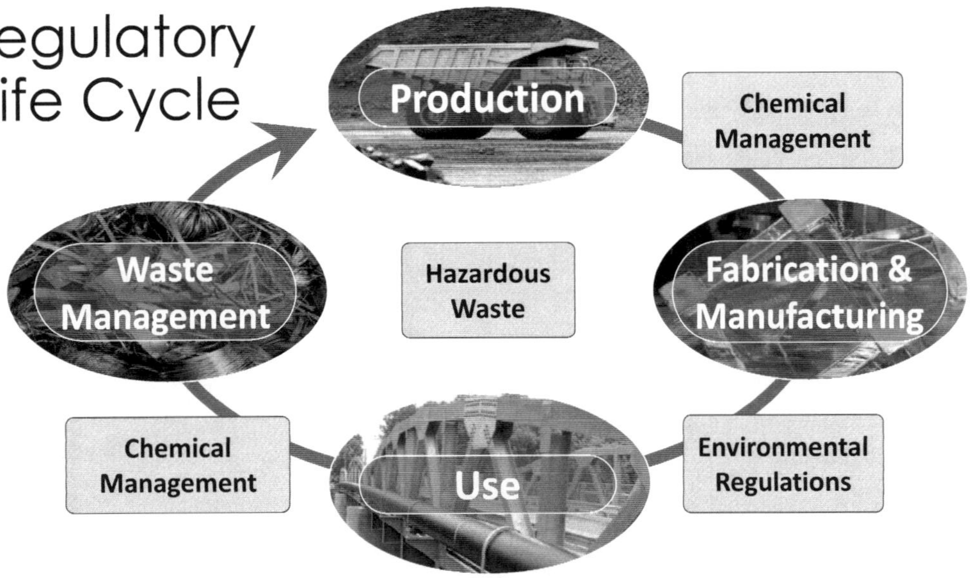

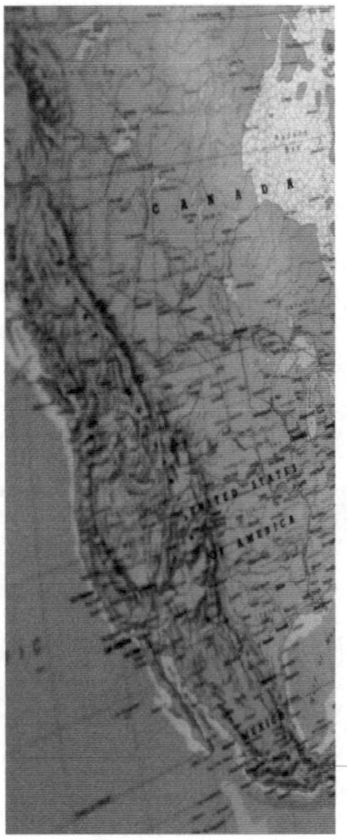

Canada

- Final Water Quality Guideline expected in 2018
  - Novel bioavailability model being applied
- Final Soil Quality Guideline expected in 2018
  - SQG would increase 20% (430 mg/kg)
- Preparing for "Zinc Group" actions under CMP
  - Assessment being prepared in 2018/2019
  - EU REACH Dossier "shared" with ECCC/HC

International Zinc Association

Chemical Management Plan – ECCC/HC –

- Phase 1 & 2 launched in 2006
  - Co (May 2017) and Se (Dec 2017) assessments and RM reports completed
- Phase 3 ongoing
  - Inorganic UVCBs not considered CEPA Toxic
  - Pb: Assessment underway, public comment in September 2017?
  - Cu, Zn: Assessment underway, public comment in March 2019
  - Cd, Fe, Mo, Ni: Assessment underway, public comment TBD
  - Al, Ge: Assessment schedule pending
- ECCC WQG updates anticipated for Zn > Cu > Ni, Pb, Fe

International Zinc Association

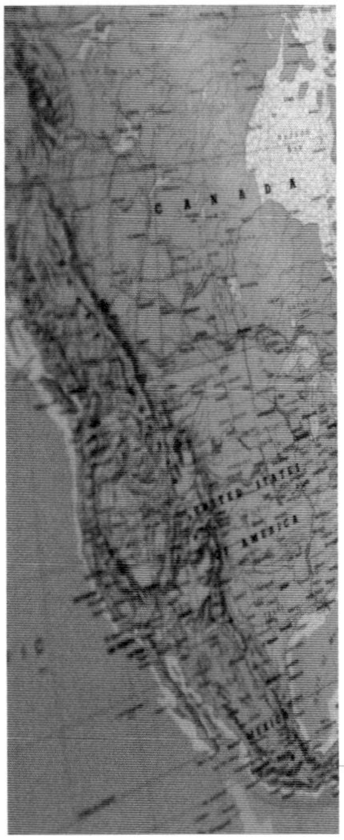

United States of America – USEPA –

- 30-yr-old water guidelines
- Promoting bioavailability-based methods (BLM)
  - 2006: Draft guideline submitted in 2006
  - 2007: Cu-BLM adopted
  - 2016: Region 5 (Great Lakes) validation studies
  - 2018: CRADA signed with 8 associations
- Establishing harmonization between standards, benchmarks and assessment practices

International Zinc Association

USEPA – Harmonized Standards

- Point-sources
  - Permanent discharges (AWQC)
  - Stormwater discharges (MSGP)
- Water Quality Assessment
  - Impaired waterbodies (303d)
  - Source reduction (TMDL)

International Zinc Association
10

USEPA – Stormwater

- Established in 1995, based on AWQC
- No substantial metals updates in 2000, 2008, or 2015
- Public comments and congressional visits
- Active involvement with National Academies committee
- Harmonized methods anticipated for 2020

USA – BLM Application for Stormwater

- IZA/AGA pilot study
- 3 to 100-fold increase!
- Initiating local petitions with operators

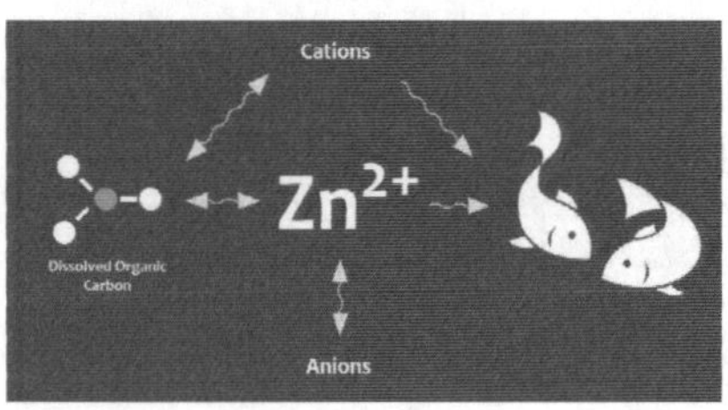

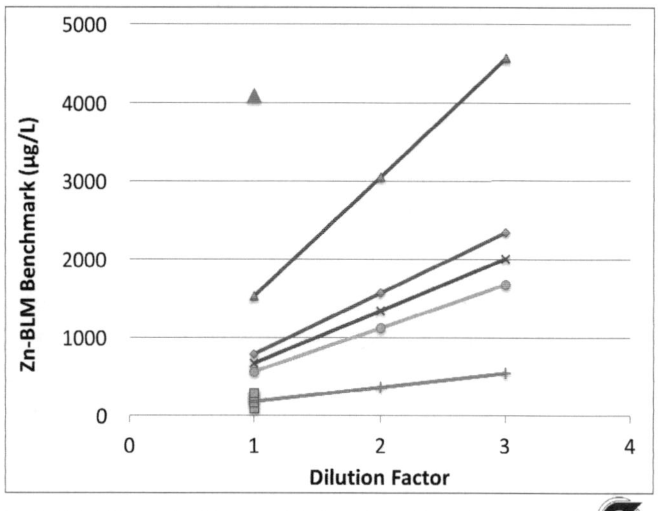

USEPA – TSCA

- Risk Evaluation (July 2017), Procedures for Prioritization (July 2017), TSCA Inventory Notification (Aug 2017)
- Comments on:
  - Mercury Inventory (Jan 2018)
  - Transparency in chemical prioritization (Jan 2018)
- Retrospective inventory notification (Feb 2018)
- Changes to CDR requirements for recycled inorganic byproducts not expected for the 2020 reporting cycle

International Zinc Association

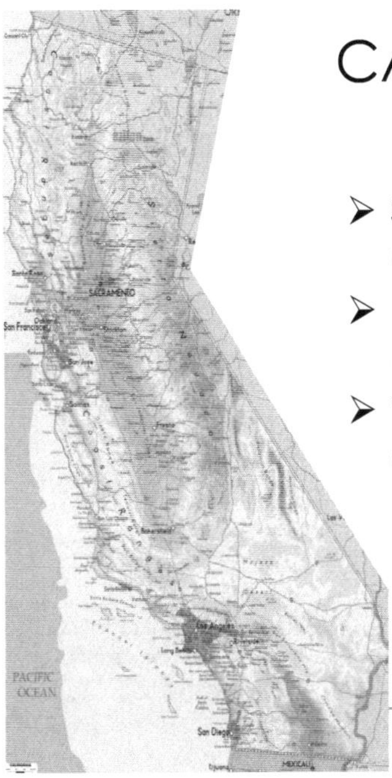

CA Attempts to Restrict Zn in Tires ...again

- State Bills in 2016 and 2017 to eliminate Zn in tires (600Kt ZnO global market)
- Bills defeated based on opposition to political overstep in scientific process
- Working on alternatives to Priority Product petition with agency, NGO, and industry partners

California Environmental Protection Agency

Department of Toxic Substances Control

California Water Boards

International Zinc Association

Latin America – Overview

- In general:
  - No formal chemical inventories in place
  - No formal pre-notification schemes
  - GHS has not been comprehensively implemented
  - Specific and localized regulations exist
- Biodiversity, clean water, and natural resources becoming moral human rights component
- Minamata (Hg) Convention: Ratified by Costa Rica (Jan), Honduras (Mar), Brazil (Aug), El Salvador (June), Argentina (Sept)

International Zinc Association

Latin America

MEXICO
- Draft Hazardous Chemicals or Mixtures In Workplaces
- Mandatory compliance by Oct 2018

COLOMBIA
- Most progressive
- GHS reaches into consumer products
- Update of labels and SDSs every three years

CHILE
- Draft Regulation on Classification, Labeling, and Notification (Nov 2017)

COSTA RICA
- 5 year transition period to comply with labeling
- Includes consumer goods

BRAZIL

CONASQ:
- registry of chemical production and imports
- risk assessment process
- risk management program

ANTT:
- Transport on highways and railways (Dec 2018)

ARGENTINA
- Draft hazardous waste regulation
- Access to potable water and sanitation are "essential human rights."

International Zinc Association

Take Away

Canada	Rapid succession of regulatory updates to support CMP Provincial/Territorial sovereignty allows flexibility
USA	Updated and harmonized federal guidelines for water State sovereignty brings uncertainty
Latin America	Progressive GHS implementation underway Mosaic of chemical management schemes anticipated

Thank You!

Eric Van Genderen
evangenderen@zinc.org
www.zinc.org
international zinc association